Ulrich Hennings

Sulfur-tolerant Natural Gas Reforming for Fuel-Cell Applications

Sulfur-tolerant Natural Gas Reforming for Fuel-Cell Applications

Sulfur-tolerant Natural Gas Reforming for Fuel-Cell Applications

by
Ulrich Hennings

Dissertation, Universität Karlsruhe (TH)
Fakultät für Chemieingenieurwesen und Verfahrenstechnik, 2009
Referenten: Prof. Dr.-Ing. Rainer Reimert, Prof. Dr.-Ing. Johannes Lercher

Impressum

Karlsruher Institut für Technologie (KIT)
KIT Scientific Publishing
Straße am Forum 2
D-76131 Karlsruhe
www.uvka.de

KIT – Universität des Landes Baden-Württemberg und nationales
Forschungszentrum in der Helmholtz-Gemeinschaft

KIT Scientific Publishing 2010
Print on Demand

ISBN 978-3-86644-459-1

Sulfur-tolerant Natural Gas Reforming for Fuel-Cell Applications

zur Erlangung des akademischen Grades eines

DOKTORS DER INGENIEURWISSENSCHAFTEN (Dr.-Ing.)

von der Fakultät für Chemieingenieurwesen und Verfahrenstechnik der
Universität Fridericiana Karlsruhe (TH)
genehmigte

DISSERTATION

von
Ulrich Hennings (M.Sc.)
aus Rottweil

Referent:	Prof. Dr.-Ing. Rainer Reimert
Korreferent:	Prof. Dr.-Ing. Johannes Lercher
Tag der mündlichen Prüfung:	24. April 2009

meiner Familie

"Man muß die Dinge so einfach wie möglich machen - aber nicht einfacher"
Albert Einstein, 1879-1955

Danksagung

Meinem Doktorvater, Herrn Prof. Dr.-Ing. Rainer Reimert danke ich sehr herzlich für die interessante und äusserst lehrreiche Zeit und die wohlwollende Förderung in allen Belangen. Herrn Prof. Dr. Johannes Lercher danke ich für die Übername des Korreferats und die wertvollen Anregungen zu meiner Arbeit. Weiterer Dank gebührt den anonymen Gutachtern der von mir verfassten Veröffentlichungen, denen ich viele wertvolle Hinweise und Anmerkungen verdanke.

Ganz herzlich bedanken möchte ich mich auch bei allen Studien- und Diplomarbeitern sowie allen wissenschaftlichen Hilfskräften, ohne deren Engagement diese Arbeit nicht möglich gewesen währe: Michael Kamm, Agustina Iturizza-Natale, Manuel Carrillo, Henrik Timmermann, Sven Siggelkow, Jose Dauden, Juan Mauel Deblas Soto, Esther de Gregorio de Miguel und Johanna Schubert.

Einen ebenso unverzichtbaren Beitrag verdanke ich den Herren Klaus Kiefer, Wilfried Stober, Frank Herter, Horst Haldenwang, Gustav Kliewer und Frau Sabina Hug, die mir bei dem Aufbau und Betrieb der Versuchsapperaturen wertvolle und tatkräftige Hilfestellung gaben. Herrn Winfried Send, dessen Engagement, Können und Geduld die Aufnahmen am Transmissions-Elektronenmikroskop möglich machte, gilt besonderer Dank.

Mcinen Kollegen und Freunden Markus Brune, Markus Wolf, Siegfried Bajohr, Frank Graf, Daniel Fouquet, André Weber, Frank Schmeer, Dominik Unruh, Martin Rohde, Hannes Storch und Marissa Maturano, sowie allen Mitarbeitern des Instituts möchte ich für die Unterstützung, angenehme Arbeitsatmosphäre am Institut, aber auch für die schönen Stunden ausserhalb der Arbeit bedanken.

Ein ganz besonderer Dank gilt meiner Familie. Meine Mutter und meine Großeltern haben immer an mich geglaubt und mit Ihrer Unterstützung diese Promotion möglich gemacht. Meiner Frau Elizabeth gilt ein liebevolles Dankeschön für Ihre unbedingte Unterstützung und vor allem für Ihr Verständnis dafür, das ich immer wieder viel der knappen Zeit zusammen für diese Arbeit aufgewendet habe.

Contents

Contents

Chapter 1.

Introduction and Objectives

The work presented is focused on fuel gas preparation for domestic scale combined heat and power generation using PEM fuel cell systems and distributed natural gas. Using a novel processing option with a novel sulfur-tolerant reforming catalyst, desulfurization of the natural gas would be made much simpler. Such a catalyst however is not readily available at this time, and therefore a large part of this work describes the preparation and characterization of such a catalyst. The remainder of this work then focusses on the catalytic properties of such a catalyst under sulfur containing feed, and finally a short economic evaluation is carried out to judge the attractiveness of the above process option.

1.1. Fuel cell systems

One intensively investigated application for fuel cells is decentralized combined heat and power generation. For this purpose, natural gas can be used as fuel, as it is often readily available through existing pipeline systems in cities. For this application, the research focuses on two fuel cell types: polymer electrolyte membrane fuel cells (PEM FC) and solid oxide fuel cells (SOFC). The SOFC concept is mainly developed by *Hexis*, while other companies such as *Viessmann*, *Vaillant*, *BBT Thermotechnik* and *Baxi innotech* develop PEM fuel cell systems. Units from *Vaillant* and *Hexis* are currently in field tests throughout Europe. Additionally, there are developments in North America (e.g. *Ballard*, *IdaTech*, *PlugPower*, *ReliOn*) and Asia (e.g. *Tokyo Gas, Ishikawajima-Harima Heavy Industries, Matsushita Electrical Industrial Co., Sanyo Electric*).

Worldwide, the 'Stationary Fuel Cell Installation Database', maintained by FuelCell2000 (www.fuelcell.org), lists 854 installed fuel cells (up and including 2008), with another 63 units listed in the planning phase. In the range of 1–5 kW, and using natural gas as fuel, a total of 141 installations are recorded. Of this total, 104 are PEM type fuel cells, 35 are SOFC fuel cells and the remainder is of unknown type. From this statistic it is clear that PEM type fuel cells are clearly favored over SOFC's, but it is also clear that other fuel cell types (such as molten carbonate- , phosphoric acid -, direct methanol - or alkali fuel cells) are not considered for decentralized combined heat and power generation.

A major feature of the SOFC in comparison to other fuel cell types is the flexibility in fuel choice. SOFCs can be operated directly on various hydrocarbon fuels without the need for a complex and cost intensive exernal fuel processing unit. The operating temperatures of SOFCs range from 650 to more than 1000 °C (to guarantee sufficient ionic transport in the electrolyte), and typically, a nickel with yttria-stabilized zirconia (Ni/YSZ) cermet is used as SOFC anode. Under the prevalent conditions, the nickel in the anode can act as an electrode but also as a catalyst, allowing the direct conversion of hydrocarbons into hydrogen and carbon (often called 'internal reforming'). On the other hand, similarly to nickel based steam reforming catalysts, unsuitable process conditions can lead to carbon deposition and deactivation of the anode. Fundamentally, from a thermodynamic equilibrium perspective, the Boudouard reaction ($2\,CO \rightleftharpoons C + CO_2$) can lead to carbon deposition. In addition, the high temperature together with surface acidity leads to cracking of higher hydrocarbons in the feed, forming so called "pyrolytic carbon". At such high temperatures elemental carbon diffuses through the nickel crystal and accumulates at crystal defects, which leads to the formation of carbon whiskers [1]. Doped ceria has attracted much attention by SOFC developers as substitute for zirconia-yttria not only because of its mixed conductivity (ionic and electronic) but also due to its lowered propensity to carbon deposition on electrodes in methane-rich atmospheres [2].

Sulfurous contaminants in natural gas do not pose a large problem in the operation of the Hexis SOFC heating appliance for the simple reason of its operating temperature at about 1000 °C. At this temperature and at the sulfur content expected in natural gas, the sulfur adsorption equilibrium lies sufficiently on the side of a clean metal surface to allow for enough residual activity. The choice for such a high temperature is driven by two reasons. First, the oxide conducting membrane performs much better at higher temperature. Second, the unit operates by autothermal reforming, and the high temperature is easily reached by mixing and combusting the anode effluent with the cathode effluent on the outer surface of a radial flow design stack. As mixing of the anode and cathode effluents at the stack outlet is desired,

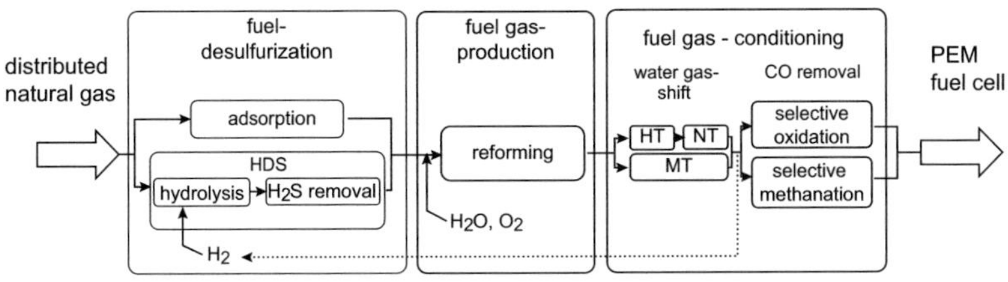

Figure 1.1: Scheme with process options for the fuel gas production of PEM fuel cell systems with a typical sulfur sensitive reforming catalyst.

this design also avoids the otherwise necessary sealing between the anode and cathode gas compartments. As mentioned above, the high operating temperature of the SOFC is driven by the much better performance of the oxygen conducting electrolyte at high temperatures. There are significant R&D efforts undertaken to develop membrane materials that offer good oxygen conducting performance at lower temperatures (500-800 °C). For these so called 'low-to-medium temperature' or 'intermediate temperature' SOFCs, the operation is envisaged to be easier due to the lower operating temperature, but sulfurous contaminants in the feed would likely become a problem in this scheme.

As a contrast to SOFCs, PEM fuel cells require pure hydrogen for operation. Thus, hydrogen must first be produced from the distributed natural gas. This, in combination with the feedgas quality requirements of the PEM fuel cell leads to a rather complex system, see Figure 1.1.

Prior to hydrogen production, sulfurous compounds must be removed from the feed stream because of their poisonous effects on the catalysts used. In industry practice, this is typically accomplished either with an absorption process (such as a "rectisol" or an amine wash) or via a two-step process, consisting of a hydrogenation of sulfur compounds to H_2S and the corresponding hydrocarbons and a following chemisorption of the H_2S, typically on zinc oxide. For fuel cell systems, one step adsorption processes were developed, but it is not clear whether the adsorbents used offer enough capacity to allow for practical service intervals [3].

After desulfurization, the natural gas gets reformed in synthesis gas. The hydrogen yield is then increased by running the syngas through one or two

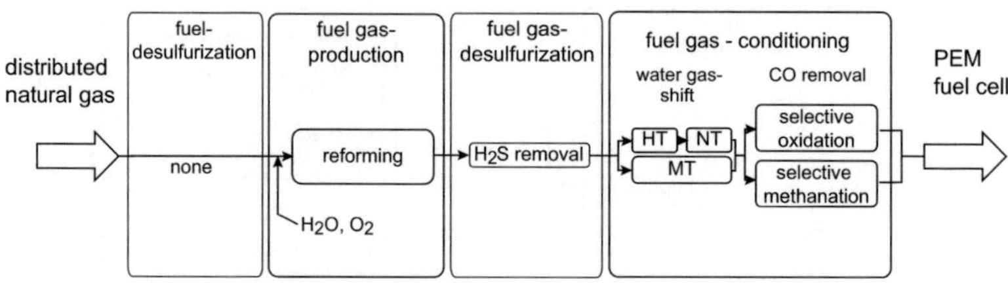

Figure 1.2: Scheme with sulfur tolerant reforming catalyst for the fuel gas production of PEM fuel cell systems.

shift steps. Following the shift steps, the residual carbon monoxide must be removed from the gas down to a very low value due to the poisonous effect of CO on the PEM anode.

In view of the relatively low sulfur content of distributed natural gas , an attractive possibility to simplify a fuel cell system would be the use of a sulfur tolerant reforming catalyst, able to convert both hydrocarbons and sulfur compounds to synthesis gas and H_2S. Hydration of the sulfur components and hydrocarbon reforming thus would be integrated in one process step and the formed H_2S could be removed downstream the reformer (see Figure 1.2). The main advantage of such a system would be a much simpler operation and faster dynamic response due to the elimination of one process step and of the external hydrogen recycle needed for the hydrogenation of the sulfur compounds. In addition, the required investment is reduced because fewer vessels, piping and control equipment are needed. The main challenge with the above described process scheme, however, is that typical steam reforming catalysts are poisoned by sulfur components.

1.2. Natural gas composition

To help in understanding the issues surrounding a sulfur-tolerant processing scheme, an overview of the composition of distributed natural gas, with an emphasis on the sulfur compounds is presented here. In principle, the composition of natural gas varies in time as well as in location, both in the hydrocarbons contained and in regards to impurities, such as sulfur compounds.

Regulation in Germany (the DVGW Arbeitsblatt G260 [4]) distinguishes between three gas families, hydrogen rich gases, methane rich gases and gases stored in liquid state. The second gas family is made up of two classes of methane rich gases, type L and type H - the main difference being the higher calorific value (H_S, often called higher heating value, HHV) and the corresponding Wobbe Index [4]. The distinction is made according to the higher heating value and Wobbe Index because this requires differently adjusted burners for type L and H gas. (Note that while in engineering practice the lower heating value, H_I is most often used because the energy released by the condensation of water is rarely utilized, in financial practice as well as in the regulation, the higher heating value is preferred, because it describes the amount of energy a user can extract from the natural gas). Especially regarding the composition, the regulation is focused on practical issues. The composition is not governed directly, but indirectly by giving limits for the hydrocarbon and water dew point, as these are of practical importance for pipeline operation. The oxygen and the sulfur contents form an exception; oxygen is limited to 0.5 vol.-% for (water-) wet and to 3 vol.-% for (water-) dry distribution systems; the sulfur content of natural gas is discussed below in more detail. Table 1.1 gives examples of the compositions of typical gases.

Before natural gas is transported, it is desulfurized, but small amounts of hydrogen sulfide (H_2S), carbon oxysulfide (COS), carbon disulfide (CS_2), and organic compounds such as sulfides (R-S-R'), disulfides (R S-S-R'), and mercaptans (R-SH) remain in the gas. From a regulation point of view, a differentiation is made between hydrogen sulfide, mercaptans and total sulfur and between the yearly average and "short-term" peaks [5]. Table 1.2 gives an overview of the regulation.

In addition to the remaining sulfur compounds from the natural gas well after purification, distributed natural gas contains organic sulfur compounds, which are deliberately added as odorants for safety reasons. In the USA, the UK and Japan, mainly isopropyl-mercaptan (IPM) or tert-butyl mercaptan (TBM) are used for this purpose (in mixture with small amounts of sulfides), while in Germany and most of Europe tetrahydrothiophene (THT) is preferred. In Germany, another DVGW Arbeitsblatt, G-280 [6], regulates the requirements for odorization.

Typical odorant loading is in the range of Y_{THT}=15–20 mg/m³ (NTP) for THT and Y_{TBM}=5–7 mg/m³ for TBM based mixtures. Assuming a base sulfur loading (for the non-odorized gas) in the range of $Y_{S,base}$=0.5–8 mg/m³, this results in a total sulfur loading of the odorized natural gas of Y_S=2.3–

Table 1.1: Typical compositions of distributed natural gas

	type H North Sea	type H Russia	type L Netherlands	type H as distributed in Karlsruhe
H_I (LHV) in kWh/m³				
	10.8	9.9	9.3	10.4
molar fraction in %				
CH_4	87.78	98.14	83.93	88.54
C_2H_6	7.34	0.61	3.63	5.52
C_3H_8	1.85	0.21	0.69	1.42
C_4H_{10}	0.52	0.07	0.25	0.48
C_5H_{12}	0.08	0.02	0.07	0.08
C_{6+}	0.06	0.01	0.08	0.06
CO_2	1.24	0.07	1.68	1.43
N_2	1.13	0.87	9.67	2.46

Table 1.2: Limits to the base sulfur loading, $Y_{S,base}$ of distributed natural gas [5]. ([†]Note: starting october 2008 this limit is not longer applicable, unless required by contractual obligations.)

	limit in mg/m³	
	yearly average	peak
mercaptans	6	16
hydrogen sulfide	5	10
total	30	150[†]

6

10.5 mg/m³ in the case of odorization with TBM and of Y_S=6–15.3 mg/m³ in the case of THT. In general, thus, a total sulfur loading in distributed natural gas of about Y_S=10–15 mg/m³ is reasonable, although short sulfur peaks of up to e.g. Y_S=150 mg/m³ are possible (see Table 1.2).

In recent years, a sulfur-free odorant has been developed and successfully introduced into the marketplace. This odorant, Gasodor^{TM} S-free^{TM}, does not add any sulfur to the gas and thus its use appears to be advantageous in combination with fuel cell systems. However, as clear from the limits to the base sulfur loading, even with no additional sulfur added due to odorization, the total sulfur content can still reach unacceptable levels from a catalyst poisoning point of view. Use of a sulfur free odorant thus does not eliminate the need for a sulfur tolerant reforming catalyst, however, it may well help to achieve sufficient activity in a sulfur tolerant catalyst.

1.3. Industrial hydrogen production

The production of hydrogen from distributed natural gas on a small scale for fuel cell applications is a recent development. The production of hydrogen from natural gas on an industrial scale on the other hand has been investigated (and practiced) for a long time. For allothermal steam reforming, nickel supported on Al_2O_3, often promoted with alkali and alkaline-earth metals, is typically used as catalyst. A plethora of literature is available on this topic, both in journals [7, 8, 9, 10, 11] and in reference books, e.g.[12, 13].

In brief, the hydrocarbon is converted with either water, oxygen or both at high temperature into synthesis gas; a mixture of mainly carbon monoxide and hydrogen. Various other chemical equilibria play a role, such as the water gas shift, the methanation, and the Boudouard reaction. The final mixture typically also contains CO_2, methane and nitrogen. If air is used instead of oxygen, the energy input for air separation is avoided, however the resulting syngas then contains a large fraction of nitrogen.

Rather than strictly adhering to the stoichiometry of the above mentioned reactions, from a catalysis point of view, it is probably more accurate to just consider several carbon, oxygen and hydrogen containing species which are adsorbed on the metal surface and which are reacting in a large number of possible reactions. In addition, especially the longer-chain hydrocarbons can be involved in several homogeneous reactions. Understanding and modelling these resulting "elementary reaction networks" is complex and laborious, e.g. [14].

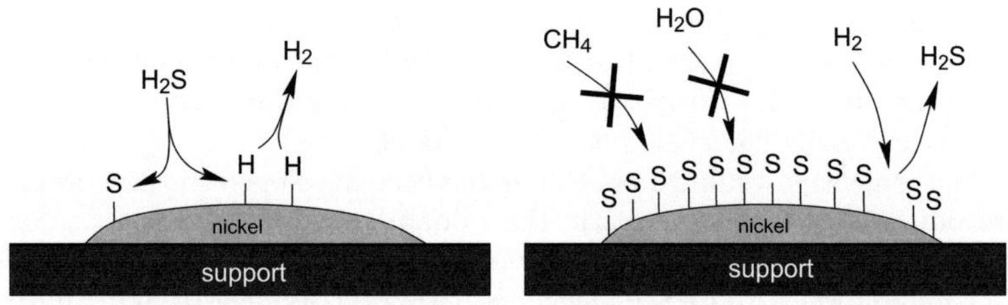

Figure 1.3: Sulfur poisoning scheme of nickel catalysts.

In both fields, hydrogen manufacturing catalysis and reactor design, research is still carried out and new discoveries are made, despite the maturity of the technology. In reactor design, especially for smaller scale systems, the heat integration of the reactor is a major challenge [15, 16, 17]. In catalysis, the main focus is on understanding and consequentially avoiding carbon deposition and sintering of both the metal and the carrier to extend catalyst lifetime [18, 19, 20].

1.4. Sulfur poisoning and sulfur tolerance

One of the main issues with industrial steam reforming - and also the issue at the heart of this work - is the sulfur poisoning of the steam reforming catalyst. Sulfur poisoning occurs because sulfur adsorbs strongly on the active metal surface area of the catalyst, forming surface sulfides [21, 1]. Sulfur is a selective poison; a partial surface coverage is sufficient for the catalyst to become essentially non-active, Figure 1.3 depicts the concept.

Sulfur poisoning has been studied in great detail for nickel catalysts. It is proposed that other transition metals are similarly affected, but as the sulfur-metal bond differs in strength for different metals, the sensitivity to sulfur poisoning differs accordingly [21, 1].

Under steam reforming conditions, the mercaptans and sulfides present in the distributed natural gas are unstable and dissociatively adsorb on the metal surface [22]. COS and CS_2 undergo hydrolysis to H_2S and CO_2 [23]. The type of sulfur compound therefore should have nearly no influence on the principle of the deactivation mechanism.

In industrial steam reformers, sulfur is adsorbed on the nickel catalyst on the top of the bed first, with an adsorption front moving through the catalyst bed. In addition, diffusion limitations also result in a profile inside the catalyst particle. Both the profiles along the bed and inside the particle depend on the ratio between the H_2S and H_2 partial pressures. Because the sulfur essentially irreversibly adsorbs on the catalyst, the operational history, together with the feed composition and the operating conditions are the determining factors for the activity of the catalyst [24, 25]. While the mechanism of sulfur poisoning is fairly well understood, the only solution practiced is the almost complete removal of all sulfur containing species from the steam reformer feed.

1.5. Ceria and ceria based materials

Mainly, the use of ceria-based materials similar to those commonly used in three way catalysts for exhaust gas treatment have been proposed to improve the reforming catalysts regarding their sensitivity to deactivation by carbon deposition [26, 27, 28]. Ahlborn et al. [29], however, also claim that the addition of ceria to noble metal catalysts increases their sulfur-tolerance in autothermal reforming.

In addition, Krumpelt et al. [30, 31] report that a platinum catalyst supported on gadolinium doped ceria showed good sulfur-tolerance. In their work on autothermal reforming of diesel for transportation applications, up to 1300 ppm of benzothiophene (weight basis) could be added to the feed without noticeable change in product gas composition. Gadolinium-doped ceria (CGO) has received much attention as electrolyte and anode material in SOFC technology (see above, also [32, 33]). The use of CGO as support material for catalysts, however, is not yet well researched.

1.6. Objectives of this work

Given the topics discussed above, the scope of this work is summarized as "Judging the feasibility of a sulfur-tolerant processing scheme for domestic sized PEM fuel cell systems", using the reports of sulfur tolerance for the noble metal catalysts supported on gadolinium doped ceria as the starting point. To achieve this goal, three main steps have to be taken:

- fabricate and characterize a catalyst similar to the one employed by Krumpelt et al.

- test and explore its catalytic properties and sulfur tolerance in the steam reforming of natural gas.

- use the gathered data to judge the feasibility of a sulfur-tolerant processing scheme.

In the following, the employed methods and experimental procedures are first introduced. Then, the "Experimental results" chapter will be concerned with the first two steps above. Finally, an additional chapter will be devoted to the practical implications and final assessment of the feasibility of the envisioned processing scheme.

Chapter 2.

Experimental methods

2.1. Catalyst preparation

All catalyst samples used in this work were prepared by glycine-nitrate combustion, according to a procedure similar to the one presented by Purohit et al. [34]. This method is illustrated in Figure 2.1. The combustion synthesis was chosen for several reasons. First, it is very easy to follow, and provides a simple one step procedure for both the support material only (as in this work) and a catalyst (with e.g. a noble metal [35]). The procedure also lends itself for transfer to a monolith preparation without further modification, where the monolith is dipped in the solution and then put in an oven. The combustion then takes place inside the monoliths, coating the walls with the desired material (e.g. [36]).

Pure ceria and gadolinium doped ceria samples were prepared by dissolving cerium nitrate ($Ce(NO_3)_3 \cdot 6\,H_2O$)), gadolinium nitrate ($Gd(NO_3)_3 \cdot 6\,H_2O$) and the stoichiometric amount of glycine (as described by [34]) in a small amount of water. The liquid was placed in a pressure-vessel. Through heating the bomb on a heating plate, the water was evaporated until a thick, viscous mass formed. In contrast to the procedure proposed by Purohit et al., who carried out the synthesis in an open beaker, the bomb was closed with a screw-on cover after evaporation of the water and further heated until ignition. After the vessel had cooled down, the resulting powder was transferred to an oven and calcined at 550 °C for 2 hours in air. See Appendix A for more details on the preparation procedure.

Similarly, the metal-containing materials were prepared by further adding the desired amount of the metal precursor (typically a metal nitrate) to the mixture. The noble metal loading ω_{Rh} was adjusted to typically 1.0 wt.-% for the reduced catalyst, however, other loadings were also prepared.

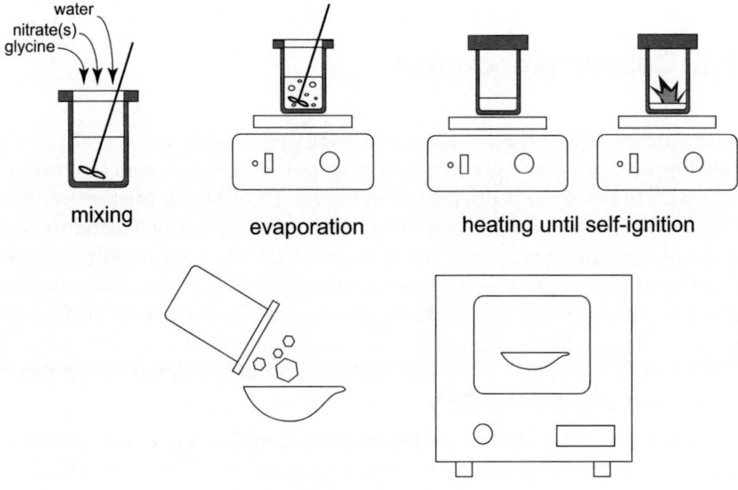

Figure 2.1: Glycine-Nitrate combustion method for sample preparation

2.2. Catalyst characterization

2.2.1. Composition, Structure and Texture

Elemental Composition   Elemental analysis was used to determine the relative amounts of ceria, gadolinium and the other elements in the solid. For bulk elemental analysis by ICP Spectroscopy, the samples were dissolved in sulfuric acid and the solution was analyzed with an Inductively Coupled Plasma Optical Emission Spectrometer (ICP-OES, Vista-Pro, Varian)[1]. In addition, both the SEM and the HR-TEM used were equipped with EDX equipment which was occasionally used (as indicated in the text).

X-ray Diffraction (XRD)   X-ray diffraction was used to yield qualitative as well as quantitative information. The XRD pattern alone can qualitatively determine the structure of the prepared material through comparison of the pattern with previously published, measured (or simulated) patterns. Also, since gadolinium ions are larger than cerium ions, their incorporation into the crystal lattice causes the unit cell to be enlarged. Quantitative analysis of the X-ray pattern yields the unit-cell size and therefore a measure of the degree of doping.

Powder X-ray diffraction (XRD) spectra were collected with a Siemens D500 diffractometer[2] in the range of $2\,\Theta = 20\text{--}80\,^\circ$ using Cu $K\alpha$ radiation, a step size of $0.01\,^\circ$ and an acquisition speed of $0.5\,^\circ/\mathrm{min}$. The program package ERACEL [37] was used to derive the unit-cell parameter from the recorded pattern.

Specific Surface Area (S_{BET})   The specific surface area (according to the BET method, S_{BET} [38]) was determined by argon adsorption using a Micromeretics ASAP 2010 unit[3] using argon at 77.35 K. A description of the method is given in [39].

Agglomerate Size   The size distribution of the agglomerates was determined by laser diffraction using a Quantachrome CILAS 1064 granulome-

---

1   Thanks to R. Sembritzki, EBI-Wasserchemie, Universität Karlsruhe (TH) for the operation of the ICP-OES
2   Thanks to B. Oetzel, Institut für Mineralogie und Geochemie, Universität Karlsruhe (TH) for the sample preparation and the operation of the diffractometer
3   Thanks to D. Schymczyk, Institut für Chemische Verfahrenstechnik, Universität Karlsruhe (TH)

ter[4]. The samples were dispersed in water/Na-polyphosphate (40 g/l) and treated with ultrasound for 800 s between measurements.

Scanning Electron Microscopy (SEM) Scanning Electron Microscopy (SEM) yields a picture of the surface topology of the crystals. In particular, the SEM picture can address the following questions: (a) particle size and particle size distribution, (b) crystal morphology (c) presence of amorphous material or other phases, and (d) in combination with EDX, the chemical composition of the present phases.

Scanning electron microscopy requires electrically electrically conducting samples, but ceria and its analogs are insulators. Samples therefore are coated with a thin layer of a platinum, and the picture obtained is in fact the picture of the thin platinum layer coating the sample.

For scanning electron microscopy (SEM), the fresh catalysts were mounted on carbon foil, coated with a thin film of platinum using a sputtering device and examined using a Leo 1530 Gemini SEM[5] operating at 3 kV.

High-resolution Transmission Electron Microscopy (HR-TEM) In contrast to scanning electron microscopy, the transmission electron microscope operates with much higher accelerating voltage and thus produces an electron beam with much higher energy. As a result, a thin enough sample is translucent to the electron beam and the resulting picture (in so called "bright field" mode) bears some similarity to an optical microscopy image.

The analogy with an optical microscope can be difficult to apply in detail, however, as for HR-TEM, the magnification can become so large that the imaged structures are in the same order of magnitude than the wavelength of the electron beam, and diffraction (rather than absorption) becomes an important mechanism in determining image contrast. The wavelength of an electron is given by the de Broglie equation. A TEM operating at 200 kV yields an electron beam with a wavelength of 0.0025 nm. Typical electromagnetic lenses further deteriorate the spatial resolution by about a factor 100, thus a spatial resolution of about 0.2 nm is achieved.

As the unit-cell dimension of a rhodium crystal is 0.38 nm and the one of ceria is 0.54 nm, the image obtained will contain some diffraction effects.

---

4 Thanks to D. Fouquet, Institut für Werkstoffe in der Elektrotechnik , Universität Karlsruhe (TH), for the operation of the instrument.

5 Thanks to P. Pfundstein, Zentrum für Elektronenmikroskopie, Universität Karlsruhe (TH) for sample preparation and operation of the instrument.

An example is visible in Figure 3.14(b) where the crystal pattern appears to continue outside the particle (so called string displacement or fringe contrast).

For high-resolution transmission electron microscopy (HR-TEM) the samples were dispersed in acetone using ultrasound and the suspension was flash-evaporated on a copper grid. The samples were examined with a Phillips CX200 microscope[6] operating at 200 kV. The microscope was equipped with an energy dispersive X-ray microanalysis system (EDX, Noran Voyager).

2.2.2. Catalysis related properties

Dispersion   The active metal surface area and from that the dispersion were determined by hydrogen chemisorption. Prior to the chemisorption experiments, the samples were prepared by heating about 50 mg of catalyst in flowing 5 % H_2 in Ar from room temperature to the desired reduction temperature T_{red}. The samples were held at T_{red} for 2 h and then cooled to room temperature in a flow of pure argon. The sample then was transferred in the chemisorption pulse-apparatus (Micromeritics Puls Chemisorb 2700). The sample was again heated under flowing 5 % H_2 in argon from room temperature to 450 °C with a heating rate of 15 K/min. After 10 min, the gas flow was switched to pure argon, the sample was heated to 500 °C, held at this temperature for 30 min to desorb hydrogen and was then cooled under Argon to room temperature.

At this point, the oven was removed and the sample was cooled to -80 °C and held at this temperature during the pulse-experiment using a methanol/dry-ice mixture. The adsorption temperature of -80 °C was selected to suppress hydrogen spillover from the noble metal to the support, which generally occurs on noble-metal/ceria samples [40].

In addition, the dispersion was determined by HR-TEM. Using several images the size of about 50 to 100 metal particles was measured. From the average noble metal particle size, the dispersion was calculated using a correlation given in [41].

Oxygen Storage Capacity (OSC) and Oxygen Exchange Capacity (OEC)   The oxygen storage capacity was determined using a thermogravi-

---

6 Thanks to W. Send, Zentrum für Elektronenmikroskopie, Universität Karlsruhe (TH) for sample preparation and operation of the instrument.

metric method similar to the one used by Stark et al. [42, 43]. The experiments were carried out with Netzsch STA 409 thermal analyser equipped with a gas pulsing apparatus. A 20 mg sample of the material was kept in flowing N_2 at 800 °C. The total OSC (the maximum amount of oxygen extractable from the sample at a given temperature) was determined by injecting H_2 pulses (1 ml) until no further weight reduction is observed. The dynamic OEC (oxygen exchange capacity, as measure of the redox kinetics) was determined by alternating between H_2 (2 ml) and O_2 (1 ml) pulses and observing the weight oscillations. Prior to both measurements, O_2 pulses were injected until no weight increase is observed to ensure the complete oxidation of the material.

Temperature Programmed Reduction (TPR) and Transient Pulse Experiments TPR was carried out by heating a 100 mg sample in a ceramic micro-reactor under flowing 5 vol.-% H_2 in argon from room temperature to 800 °C with a heating rate of 15 K min^{-1}. The effluent gases were monitored using a mass-spectrometer (MKS Minilab). The TPR experiment was operated such that a large H_2 surplus avoids hydrogen profiles along the catalyst. This results in a largely invariant hydrogen partial pressure and requires the measurement of the reaction product H_2O to establish the TPR curves.

In the same apparatus, pulse experiments were carried out by flowing an inert gas over the catalyst (Ar was used) and injecting pulses of reactive gases (e.g. CH_4, CO_2, H_2S) or water with a syringe through a rubber septum (typical pulse was 41.5 μmol). The effluent gas composition and its time-dependence was monitored using a mass-spectrometer (MKS Minilab)[7]. The mass-spectrometer was connected directly at the reactor outlet using a heated PTFE capillary. Figure 3.5 shows a flow scheme of the employed apparatus. The mass spectrometer distinguishes between ions of different mass to charge ratio (m/z). To relate the mass to charge ratio m/z to the species to be measured, the fragmentation pattern of the species molecule must be known. The m/z values used for measuring the species of interest here are shown in Table 2.1.

---

7   Thanks to the Institut für Werkstoffe der Elektrotechnik for the generous loan of the instrument

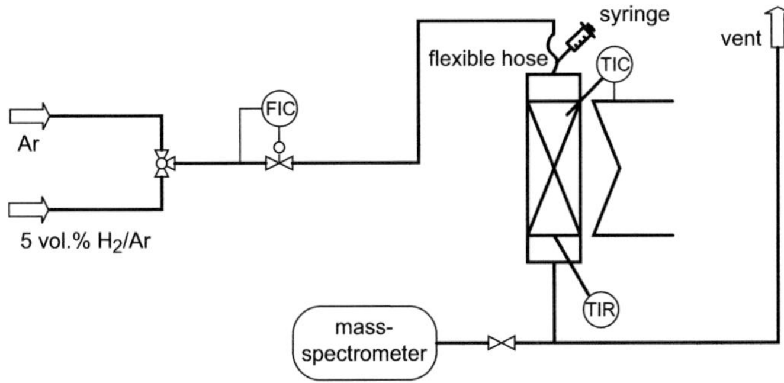

Figure 2.2: Apparatus used for TPR and transient pulse experiments

Table 2.1: mass to charge ratios m/z used for measuring with the mass-spectrometer

component	ion	m/z value
CH_4	CH_3^+	15
CO	CO^+	28
H_2	H_2^+	2
CO_2	CO_2^+	44
H_2O	H_2O^+	18
N_2	N_2^+	28
H_2S	H_2S^+	34
CS_2	CS^+	76
C_2H_6S (Ethylmercaptan, EM)	EM^+	62
i-C_3H_8S (iso-Propylmercaptan, IPM)	IPM^+	76
tert-$C_4H_{10}S$ (tert-Butylmercaptan, TBM)	TBM^+	90
C_4H_8S (Tetrahydrothiophen, THT)	THT^+	88

Table 2.2: Composition of the model natural gas used for the experiments (natural gas type H, see chapter 1.6)

component	volume fraction y_i in %
CH_4	85.70
C_2H_6	5.20
C_3H_8	2.00
n-C_4H_{10}	0.30
i-C_4H_{10}	0.18
n-C_5H_{12}	0.01
CO_2	1.51
N_2	4.98

2.3. Reforming Experiments

2.3.1. Steam reforming apparatus

The steam reforming reaction was conducted in a fixed-bed ceramic reactor with 7 mm inner diameter (d_R). Three-hundred mg of the fresh catalyst were diluted with about 2000 mg quartz, and the mixture was placed in the reactor supported on a quartz wool plug. The reactor was equipped with a thermocouple to monitor the temperature at the outlet of the catalyst bed. The reactor was heated with an electric oven; the gases were dosed using mass flow controllers. Liquid water was dosed with a mass flow controller and was evaporated with a falling film vaporizer.

Prior to the reaction experiment, the catalysts were reduced in a flow of 20 vol.-% H_2 in N_2 at 500 °C for 3 hours. Then the reactor was heated to 800 °C under the same flowing gas, and held at this temperature for 1.5 hours. To stabilize the catalyst's activity, the gas composition was cycled between this gas and synthetic air (20 vol.-% O_2 in N_2) at 800 °C 5 times with holding times of 1.5 hours and a 5 min N_2 purge in between each cycle. The above procedure is a result of the experiments detailed in Section 3.1.2 .

As feedgas either pure methane, methane-nitrogen mixtures or a model natural gas was used (see Table 2.2). The steam-to-carbon ratio (S/C ratio ζ) is defined as the ratio of water molecules to carbon atoms, ${}^N\Phi_{H_2O}:{}^N\Phi_C$, flowing into the reactor. The reactor effluent is cooled in two steps, first in a counter current gas cooler with cooling water and second in a refridgerator-type gas cooler set at -4 °C.

The (dry) gas composition was monitored using a Varian CP-2003 Micro-GC equipped with a molsieve 5A, a PoraPlotQ column and thermal conductivity detectors; helium was used as carrier gas. Both the plant and the Micro-GC were controlled by a PC with a self-written LabView program. A flow scheme of the steam reforming apparatus is shown in Figure 2.3

2.3.2. Evaluation of the experiments

Experimentally, the inlet flow and composition, reactor (outlet) temperature and pressure (both at the inlet and the outlet) and the composition at the outlet are the measured quantities (see Figure 2.3, note that the dry gas composition is measured). Since the steam reforming reaction increases the number of molecules, the volumetric flow rate is changed by the reaction. For calculation of the volumetric flowrate at the reactor-outlet, two methods can be chosen. First, nitrogen can be used as internal standard. With a known gas composition at both the inlet and outlet, and assuming that the nitrogen does not react under steam reforming conditions (see Appendix F for validation), the outlet volumetric flow can be calculated as

$$^V\Phi_{out} = {^V}\Phi_{in} \frac{y_{N_2,in}}{y_{N_2,out}} \tag{2.1}$$

With the volumetric flow at the outlet and the composition known, the molar rates of all species can be calculated at both the in- and the outlet, enabling the calculation of e.g. the methane conversion:

$$X_{CH_4} = \frac{^N\Phi_{CH_4,in} - {^N}\Phi_{CH_4}}{^N\Phi_{CH_4,in}} \tag{2.2}$$

and the selectivity towards CO_2 (for the case of methane being the only carbon containing species in the feed):

$$S_{CO_2} = \frac{^N\Phi_{CO_2}}{^N\Phi_{CH_4,in} - {^N}\Phi_{CH_4}} \tag{2.3}$$

The internal standard method has one drawback, however. Typically, the nitrogen content of natural gas is rather small (see Table 2.2), and the volume expansion further reduces the nitrogen content. This leads to a

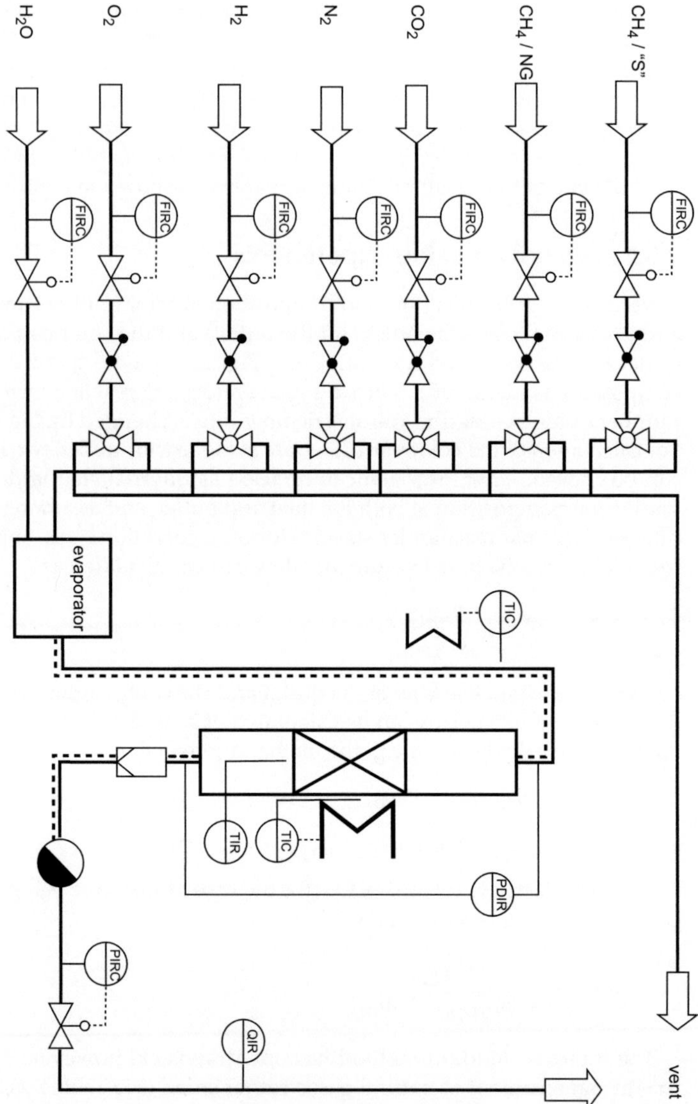

Figure 2.3: Apparatus used for steam reforming experiments

rather large uncertainty in the measured nitrogen content, which in turn leads to a large uncertainty in the calculated conversion and selectivity.

A second method is to assume that no carbon deposition takes place in the reactor. Then, the correct volumetric flow at the outlet of the reactor can be determined by a carbon balance. This has experimentally the advantage that the measurement of the carbon containing species is more accurate, leading to more accurate results. In the case of the catalysts investigated here, the assumption that no carbon deposition takes place in the reactor is correct in most cases (see Section 3.3.1), the latter method was therefore employed.

The difference between the calculated nitrogen content and the experimentally determined value as well as the measured pressure drop in the reactor were used as indicators for the validity of the no carbon deposits assumption.

Species balance The equation of continuity for each species i and several chemical reactions j is written [44]:

$$\frac{\partial}{\partial t} c_i = -\left[\nabla \cdot c_i \vec{u}\right] - \nabla\left[-D_{ii'} \nabla c_i\right] + \sum_{all\ j} \frac{\nu_{i,j} r_j}{\varepsilon} \tag{2.4}$$

with c_i, the concentration of species i being defined as

$$c_i = \frac{n_i}{V} \tag{2.5}$$

i.e. the amount of species i per unit gas volume, and r_j, the reaction rate of reaction j being defined as

$$r_j = \frac{1}{V_R} \frac{1}{\nu_{i,j}} \frac{dn_i}{dt} \tag{2.6}$$

i.e. the amount of species i reacting per unit time within the reactor volume. Note that the reference volumes are different - the concentration is related to the gas volume only, while the reaction rate is related to the reactor volume, i.e. the volume of the catalyst bed including both catalyst particles and the void space between them. Both volumes are related through the void fraction of the bed ε, which is defined as

$$\varepsilon = \frac{V}{V_R} \tag{2.7}$$

Only accounting for steady state and neglecting both diffusion and dispersion, Equation 2.4 simplifies to:

$$[\nabla \cdot c_i \vec{u}] = \sum_{k=1}^{n} \frac{v_{i,j} r_j}{\varepsilon} \tag{2.8}$$

(Note: the disregard of diffusion and dispersion effects is justified for the chosen experimental conditions - see Appendix D)

Considering only the axial concentration gradients yields:

$$\frac{d(c_i u_z)}{dz} = \sum_{j=1}^{n} \frac{v_{i,j} r_j}{\varepsilon} \tag{2.9}$$

It is more practical to use the the superficial velocity u_0, which is related to the interstitial velocity u_z by $u_0 = u_z \varepsilon$. Further using the cross-sectional area of the reactor (without packing, assumed to be constant along the length of the reactor and constant over time), A_R, Equation 2.9 becomes:

$$\frac{d(^{N}\Phi_i)}{dz} = A_R \varepsilon \sum_{j=1}^{n} \frac{v_{i,j} r_j}{\varepsilon} = A_R \sum_{j=1}^{n} v_{i,j} r_j \tag{2.10}$$

with $^{N}\Phi_i$ being the (molar) flow of species i.

Momentum Balance   The steam reforming reaction is highly endothermal, and to avoid temperature gradients in the reactor the catalyst must be diluted. In addition, the fabricated catalyst has a rather small particle diameter (ca. $15\,\mu$m, see Table 3.1). Both these factors lead to a rather large pressure drop across the reactor, which has to be considered in the model. In general, the momentum balance can be formulated as [44]:

$$\frac{\partial}{\partial t} \rho \vec{u} = -[\nabla \cdot \rho \vec{u} \vec{u}] - \nabla p - [\nabla \cdot (-p\delta + \Gamma)] + \rho g \tag{2.11}$$

Only stationary flow conditions are considered, thus the accumulation term on the left hand side can be neglected. Assuming that no radial pressure gradients are present, neglecting the influence of gravity and using the friction factor approach, Equation 2.11 simplifies to:

$$-\frac{dp}{dz} = \frac{\rho u_0^2}{d_p} f \tag{2.12}$$

A multitude of representations for the friction factor, f can be found in literature (e.g. [45, 46]); Ergun's equation (Equation 2.13) is used most commonly.

$$f = \frac{(1 - \varepsilon)}{\varepsilon^3} \left[\frac{150(1 - \varepsilon)}{Re_p} + 1{,}75 \right] \tag{2.13}$$

In Equation 2.13 Re_p is the Reynolds number related to the particle diameter, $Re_p = \frac{u_0 d_p \rho}{\mu}$, in which d_p is the catalyst particle diameter, u_0 the superficial velocity (the velocity of the fluid if there were no packing present), ρ is the gas density and μ the dynamic viscosity.

Heat and mass transport   For the conditions employed in the experiments, temperature gradients as well as (both external and internal) mass transfer effects can be neglected. See Appendix D for more details.

Rate expressions for the chemical reactions   To solve the set of continuity equations (Equation 2.9), reaction rate expressions are needed. Details regarding the reaction rate expressions are given in section 3.5

2.3.3. Model implementation

Full model   The above equations were implemented in "Gnu Octave", a software package freely available under GNU licence and primarily intended for numerical computations (see http://www.octave.org). Octave contains an implementation of the DAE solver DASPK (developed by Brown, Hindmarsh and Petzold [47]), which was used for the integration of the differential equation system. For parameter estimation, the also freely available package "leastsqr" (written by Richard Shrager, Arthur Jutan and Ray Muzic), which employs a Levenberg-Marquardt nonlinear regression was used.

The required thermodynamic data set for the water gas shift equilibrium was extracted from the database of the process simulation package ASPEN Engineering Suite, the dynamic viscosity of the mixture was determined by a method given by Reid et al. [48].

Simplified model   In section 3.5, it is shown that the reaction rate can approximately be described as first order in methane and zeroth order in water (and therefore $r_1 = k_1(c_{CH_4})^1$) with acceptable error (see Table 3.7, case 4 and accompanying text). This insight was used to develop a much simpler model. Considering only the methane conversion and approximating the pressure change by using a mean reactor pressure $\bar{p}$, using the experimentally determined values p_{out} and Δp:

$$\bar{p} = p_{out} + 0.5\Delta p \tag{2.14}$$

Equation 2.10 simplifies to

$$dX_{CH_4} = \frac{-A_R}{{}^N\Phi_{CH_4,in}} r_{CH_4} dz \tag{2.15}$$

Note that in Equation 2.15, r_{CH_4} is used instead of r_1, which would be correct according to the nomenclature used above. As in the simplifies treatment only reaction 1, i.e. methane conversion by steam reforming is considered, the more descriptive r_{CH_4} (and also k_{CH_4}) is used from now on for the value derived from the simplified method described above.

To account for the increase in the number of species by the reaction the fractional increase Ψ is defined as

$$\Psi = \frac{{}^N\Phi_{out}(X_{CH_4} = 1) - {}^N\Phi_{in}}{{}^N\Phi_{in}} \tag{2.16}$$

where ${}^N\Phi_{out}(X_{CH4}{=}1)$ signifies the total molar flow rate exiting the reactor at full methane conversion and ${}^N\Phi_{in}$ signifies the total molar flow rate entering the reactor. In the case of natural gas as feedgas, the volume expansion is also depending on the conversion of the other hydrocarbons and the inert content of the natural gas used. For the natural gas, the following definition was used:

$$\Psi = \frac{{}^N\Phi_{out}(X_{"all\ hydrocarbons"} = 1) - {}^N\Phi_{in}}{{}^N\Phi_{in}} \tag{2.17}$$

It should be noted, however, that the difference in the resulting Ψ is only minor for the natural gas composition used (see Table 2.3).

Using Ψ, the integrated form of Equation 2.15 can be written as ([49])

$$k'_{CH_4}\tau = -(1 - \Psi)ln(1 - X_{CH_4}) - \Psi X_{CH_4} \tag{2.18}$$

where τ is the residence time, defined as

$$\tau = \frac{V_R}{{}^V\Phi_{in}} \tag{2.19}$$

Table 2.3: Factor Ψ as function of the steam to carbon ratio ζ and the feedgas type (NG evaluated for the composition given in 2.2)

ζ	CH$_4$ only	NG
1.5	0.800	0.813
2.0	0.667	0.676
2.5	0.571	0.578
3.0	0.500	0.505
3.5	0.444	0.448
4.0	0.400	0.403
5.0	0.333	0.336

In Equation 2.19, V_R is the reactor volume and $^V\Phi_{in}$ the volumetric flowrate at reaction conditions entering the reactor.

Alternatively, the modified residence time τ_{mod} defined as the ratio of the catalyst's mass and the volumetric flow rate at the inlet:

$$\tau_{mod} = \frac{m_{cat}}{^V\Phi_{in}} = \rho_{bed}\tau \tag{2.20}$$

can be used. The modified residence time is related to the residence time by the catalyst's bed density $\rho_{bed} = \frac{m_{cat}}{V_R}$, and consequently the rate constant becomes

$$k'_{CH_4}\tau = \left(\frac{1}{\rho_{bed}}k'_{CH_4}\right)\tau_{mod} = k_{CH_4}\tau_{mod} \tag{2.21}$$

Due to the effects of deactivation, the effective rate constant k'_{CH_4} is not constant, but depends on the degree of deactivation and thus may be a function of the operating conditions, the feedgas composition (particularly the amount of poison) and also of time-on-stream.

The rate constant k_{CH_4} (measured at temperature T and pressure p) can be converted into the turnover frequency (TOF) by accounting for the methane volume fraction y_{CH_4} in the feed and the active sites on the catalyst:

$$TOF_{CH_4} = k_{CH_4}(T,p)\frac{y_{CH_4}}{V_m(T,p)}\frac{M_{Rh}}{\omega_{Rh}D_{Rh}} \tag{2.22}$$

Chapter 3.

Experimental results

For a preliminary study, rhodium, ruthenium and platinum catalysts supported on ceria and gadolinium doped ceria were prepared and evaluated. The results of this comparison are published in a research paper [35]. Based on this comparison, it was decided to further investigate catalysts with rhodium as noble metal. In addition to the activity and the sulfur tolerance of the catalysts, two main topics warranting further and more in-depth investigation were identified, namely the redox properties of the support and the stability of the catalyst.

Redox properties of the support One important feature of catalysts supported on ceria and related materials is that they easily undergo redox cycles at temperatures above 600 °C. Further, these redox properties can be fine-tuned to the application by substituting a fraction of the Ce^{4+} ions in the ceria crystal by other ions. For example, in automotive emission reduction, the newer generation three way catalyst (TWC) contains zirconia-doped ceria because this material has an enhanced reducibility (in this application called oxygen storage capacity, OSC) compared to pure ceria.

The property to store and release oxygen would also be attractive for hydrogen production by steam reforming of natural gas, as one of the typical problems is the degradation of the catalysts by formation of carbon deposits (see Section 3.3.1).

Also, the poisoning effects of sulfur on transition metal catalysts are well known (see Section 1.6). In addition, the OSC of ceria containing catalysts is deactivated by SO_2, usually present in flue gas due to sulfur-impurities contained in hydrocarbon fuels. The OSC is deactivated because sulfur reacts with ceria, forming various Ce-S-O species [50]. While the interaction between sulfur and ceria is rather complex and therefore not fully understood yet, it is clear that the redox properties of the materials largely influence the sulfur adsorption properties [51].

While the redox properties of ceria and ceria zirconia catalysts are well characterized, the properties of gadolinium doped ceria are not as well documented so far. Thus, several experiments were conducted to characterize the structure and the redox properties of the support material.

Stability of the catalyst   The stability under reaction conditions was poor in the preliminary study [35], so various experiments were conducted to understand the deactivation and improve the stability.

Note that the results of the investigations regarding the catalyst stability and the redox properties presented here were also published in two research papers [52] and [53].

3.1. Catalyst preparation and characterization

3.1.1. Structure, texture and redox properties of the support

Structure of the support   Analyzing the recorded XRD spectra (see Figure 3.1, results shown for the samples calcined at 550 °C only), the segregation of a second phase can be observed starting at a gadolinum content of $x_{Gd} = 0.35$, indicating the solubility limit of the gadolinium ion in the ceria lattice. For the samples with $x_{Gd} < 0.35$ no other peaks can be observed, indicating a complete combustion and the absence of any residues. Figure 3.2 shows the XRD spectra of a sample with $x_{Gd} = 0.3$ calcined at several temperatures. No formation of a second phase can be discerned with increasing heat treatment; the material thus appears to be stable even at temperatures up to 1100 °C.

The lattice parameter of the doped ceria phase was refined from the XRD spectra using the computer program ERACEL [37]. The corresponding plot of the refined lattice parameter vs. the molar gadolinium content x_{Gd} is shown in Figure 3.3.

As expected, the unit cell increases with x_{Gd}, indicating the integration of the larger gadolinium ion into the fluorite lattice. Moreover, no significant difference can be seen for the two different calcination temperatures. Note that using ICP Spectroscopy it was shown that the elemental composition of the synthesized catalysts is very close to the desired composition (see Appendix B).

The lattice parameter only increases up to $x_{Gd}{=}0.3$, which is in line with the occurrence of a segregated phase (see above) and also with previous

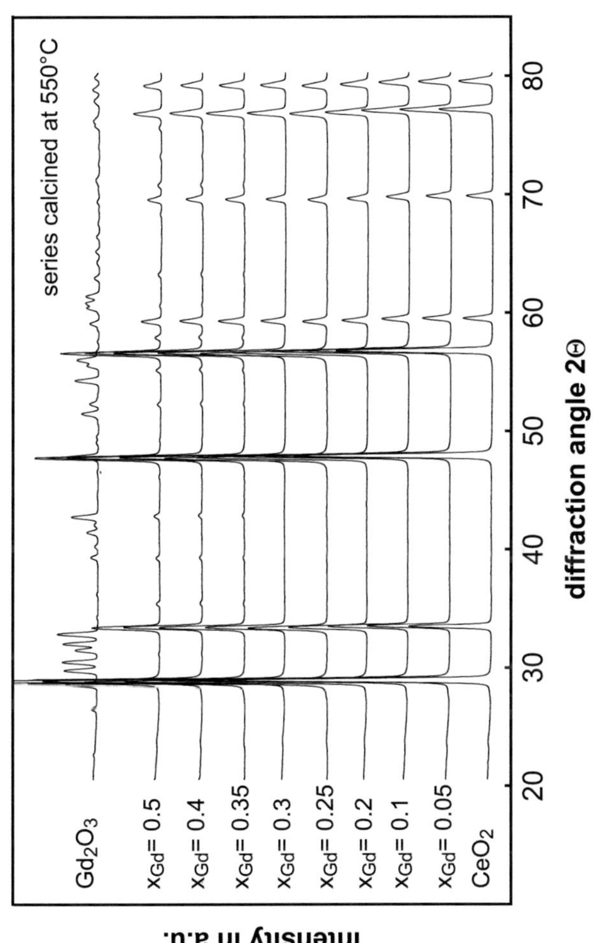

Figure 3.1: XRD spectra of the samples indicating segregation of a separate phase at gadolinum contents of $x_{Gd} = 0.35$ and higher.

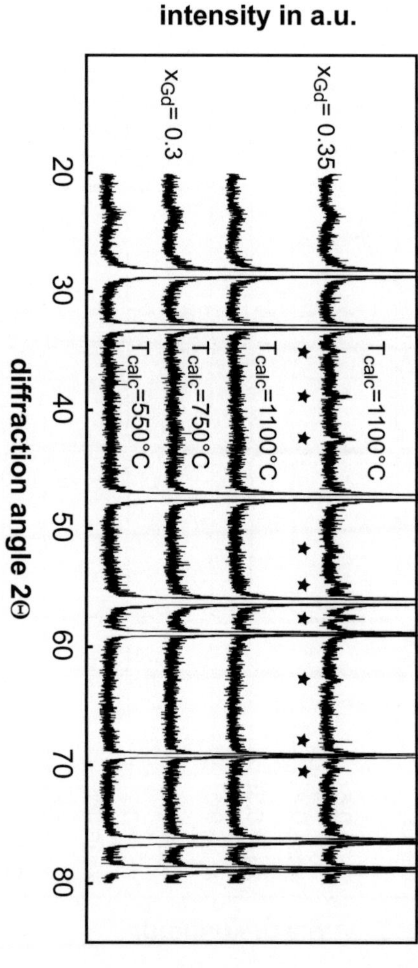

Figure 3.2: XRD spectra of the samples with a gadolinum content of x_{Gd} = 0.35. For comparison, the spectrum of the sample with x_{Gd} = 0.35, calcined at 1100 °C is shown along with the marked positions of the reflections of the second phase.

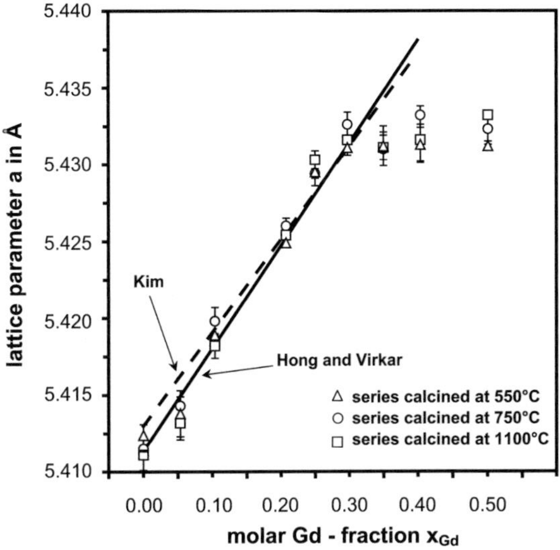

Figure 3.3: Lattice parameter calculated from XRD spectra. The error bars correspond to the standard deviations obtained by the data fitting. The lines correspond to the unit cell predicted by Kim's and by Hong and Virkar's model, respectively.

studies [54, 55] Kim [56] as well as Hong and Virkar [55] published an empirical relation between the ionic radii of the ions and the lattice parameter of fluorite type oxides; both correlations are included in Figure 3.3.

Texture of the support The morphology of the synthesized materials can be seen in Figures 3.4(a)-(d). The obtained material is a large agglomerate of many small sintered crystallites, the particles appear smooth on the surface. As a first approach and judging from the visual appearance, neither the different gadolinium content nor the calcination temperature appears to introduce large changes.

The measured BET surface is given in Table 3.1. As could have been expected from the electron microscopy images, the specific surface area of all

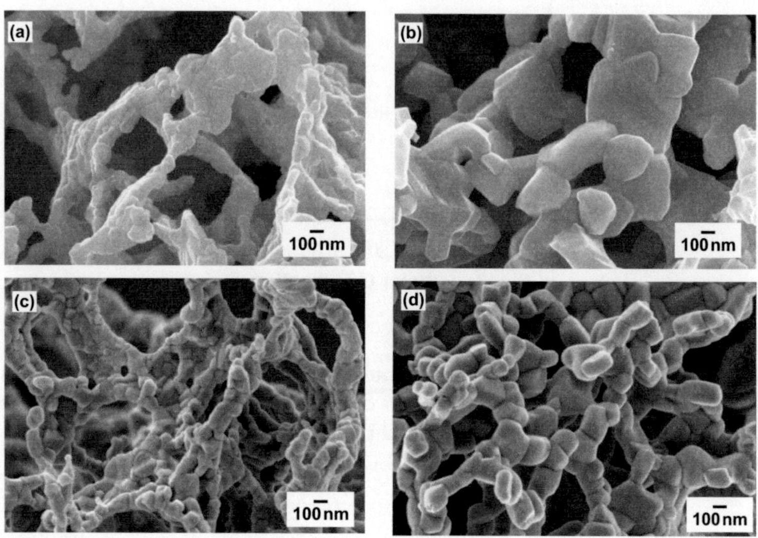

Figure 3.4: SEM images of CeO_2, calcined at (a) 550 °C and (b) 1100 °C and $Ce_{0.8}Gd_{0.2}O_{1.9-\delta}$, calcined at (c) 550 °C and (d) 1100 °C

Table 3.1: Textural properties. The $d_{p,A}$ values in parentheses represent the measured diameter after an ultrasonic treatment, see text.

x_{Gd}	S_{BET} in m²/g		d_p in nm		d_A in µm	
T_{calc} in °C	550	1100	550	1100	550	1100
CeO_2	3.7	2.8	183	175	11.88 (7.40)	13.49 (8.75)
x_Gd=0.1	4.2	3.7	180	179	10.84 (6.08)	12.54 (8.50)
x_Gd=0.2	3.8	3.4	182	179	8.84 (4.83)	13.75 (8.73)
x_Gd=0.3	4.0	3.1	174	187	12.50 (7.64)	13.85 (9.77)

prepared samples is rather low, which implies that there is no microporous structure. The variation in surface area with x_{Gd} is only small but it can be noted that the samples calcined at 1100 °C always show a lower specific area compared to those calcined at 550 °C, probably due to some sintering.

Low specific surface areas are not uncommon for ceria. Ramírez-Cabrera et al. report a specific surface area for ceria calcined at 1000 °C of 2 m²/g [57]. In another paper by Aneggi et al., a series of ceria samples was calcined at different temperatures, yielding 57–123 m²/g at 550 °C, but 4–10 m²/g at 800 °C, depending on the preparation method [58]. Similarly, Stark et al. were able to synthesize zirconium doped ceria supported platinum catalysts with a specific surface area of 78–98 m²/g when calcined at 700 °C [43]. However, calcination at 1100 °C also lead to a specific surface area of about 3.5–4.5 m²/g for their samples [43].

During the catalyst preparation, the material is expected to be exposed to temperatures around 1200 °C, albeit only for a very short time (see Appendix A). Nevertheless, the preparation method itself must be responsible for the observed low surface area and given the intended application as catalyst support, the low surface area of the synthesized materials is unfavorable. However, as is clear by the above cited literature examples, high surface area ceria materials are prone to sintering at temperatures in the range of the desired operating temperature (ca. 800 °C). As temperature deviations during startup, shutdown or operation could lead temporarily to even higher temperatures in the reactor, a surface area close to the samples examined here may be the best achievable result.

From the SEM images, the average crystal size d_p was obtained (see Table 3.1). There is no systematic variation of primary particle size with either composition or calcination temperature. Apparently, no particle growth is induced by the applied heat treatment. Also, the composition neither seems to have any influence on particle size nor on sintering propensity.

The average agglomerate diameter d_A is also given in Table 3.1. The values in parentheses represent the measured diameter after an ultrasonic treatment of 800 s length. For the samples calcined at 550 °C, the diameter decreases because of the ultrasonic treatment to about 55–60% of the initial value, indicating that the calcination treatment at this temperature does not fuse the primary particles together. The initial diameters for the samples calcined at 1100 °C are larger and decrease even slightly more (to 60–70%).

The material thus consists of rather soft agglomerates, and the calcination has no effect on the agglomerate hardness (i.e. the calcination treatment does not fuse particles together). In conclusion, the material is stable in the temperature range expected for the steam reforming reaction.

Redox properties of the support The TPR patterns shown in Figure 3.5 give insight into the redox properties of the materials calcined at 1100 °C. As clearly seen in Figure 3.5, there is extensive water formation at temperatures above 600 °C, indicating that Ce^{4+} is reduced to Ce^{3+} in pure ceria.

For the doped samples, the onset of water formation generally shifts to lower temperatures, while the peak water formation is not quite as intense. This trend only holds true for the samples with lower gadolinium contents, however. The x_{Gd}=0.3 sample exhibits a higher onset temperature compared the x_{Gd}=0.25 sample.

The peak at high temperature seen most strongly for pure ceria is usually interpreted to be due to bulk reduction, a process that is regarded by most authors not to be limited by the reaction kinetics but by the oxygen diffusion in the bulk of the material [59]. Although experimental values for the diffusion coefficient of oxygen in ceria differ strongly, it is generally agreed that the activation energy for diffusion is lower for doped than for pure ceria [60]. It is thus understandable why the reduction onset is at lower temperatures for the doped samples.

Some authors disagree with the above interpretation [59]. They attribute the onset at lower temperatures to effects related to the surface reaction, e.g. the activation energy to create surface defects, which is lower for doped ceria [60]. In any regard, both hypotheses (which were formulated for other ceria materials) can explain the behavior observed for the Gd-doped samples.

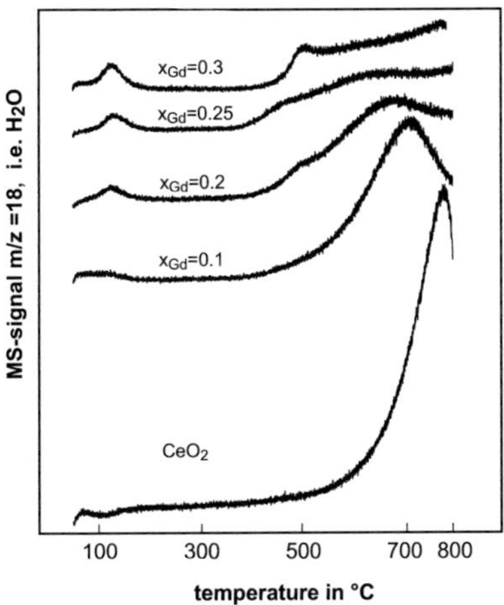

Figure 3.5: TPR patterns of the fresh samples calcined at $1100\,^\circ$C.

In gadolinium doped ceria, oxygen vacancies are present due to the doping, and additional ones are formed upon reduction. In Kröger-Vink notation, [61] the formation of a solid solution (i.e. doping) and the reduction process of the ceria can be described as

$$(1-x)CeO_2 + xGd_2O_3 \rightleftharpoons$$
$$(1-x)Ce_{Ce} + xGd_{Ce} + x/2V_O^{\bullet\bullet} \tag{3.1}$$
$$Ce_{Ce} + 2O_O + \delta/2H_2 \rightleftharpoons$$
$$Ce'_{Ce} + (2-\delta)O_O + \delta V_O^{\bullet\bullet} + \delta/2H_2O \tag{3.2}$$

Ce_{Ce}:   ceria ion on a ceria lattice position
Gd_{Ce}:   gadolinium ion on a ceria lattice position
O_O: oxygen ion on a oxygen lattice position
$V_O^{\bullet\bullet}$:   2-fold positive vacancy on a oxygen lattice position, i.e. O^{2-} removed from lattice
Ce'_{Ce}:   1-fold negatively carged ceria ion on a ceria lattice position, i.e. Ce^{3+} instead of Ce^{4+}

The presence of vacancies due to the doping will result in a lower degree of reduction compared to ceria at the same conditions, because the equilibrium is shifted towards the left in Equation 3.2. The flatter reduction peak for the doped samples thus can be rationalized.

35

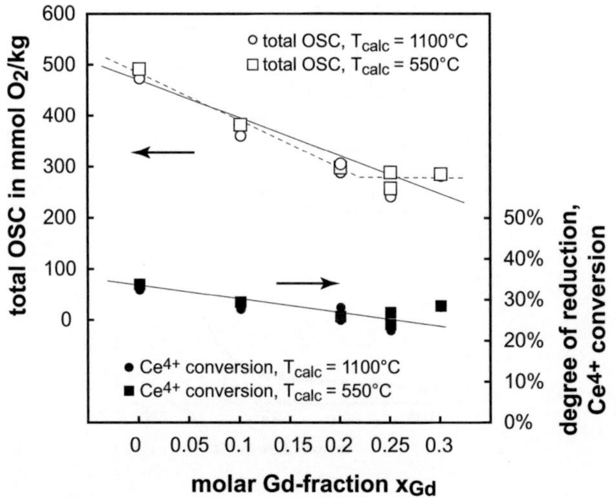

Figure 3.6: Total oxygen storage capacity and corresponding degree of reduction at 800 °C. See text for the discussion of the grey lines.

The peak seen in the low temperature region around 100–150 °C for the samples is commonly interpreted as the reduction of surface species. Obviously, the doping results in an increased reducibility at this low temperature, or differently put, lowers the activation barrier for the reduction. In addition to experimental studies that showed a similar trend in the low-temperature reduction peak for ceria-zirconia catalysts [59], a computational study using a force-field approach also predicted a lower activation energy for reduction of gadolinium doped ceria compared to pure ceria [62].

The redox properties of the materials were further studied at 800 °C by means of reduction in a TG-setup, which allowed for quantification of the degree of reduction. The temperature was chosen because it is the preferred operating temperature for steam-reforming.

Figure 3.6 shows the total oxygen storage capacity (OSC), i.e. the maximum amount of oxygen extractable from the sample at this temperature (see Appendix C for details on the measuring method).

As expected, the total OSC drops dramatically with the incorporation of gadolinium ions in the lattice, but levels off for higher gadolinium levels (see Figure 3.6). These experiments therefore confirm the qualitative observation in the TPR patterns for the high temperature peak. Further, no significant difference can be seen for the samples calcined at the two temperatures. This is likely due to the low specific surface area of the samples.

One interesting feature is that the reducibility was measured to be roughly the same for all samples with $x_{Gd} \geq 0.2$. Nevertheless, one can interpret a trend in the measurements (as indicated by the solid grey line in Figure 3.6), with the deviations attributed to experimental error. This corresponds with the equilibrium argument made above that more gadolinium incorporation hinders the creation of more vacancies, which leads to lower reducibility. An alternative explanation, corresponding to the dashed gray line will be detailed later in the discussion of Table 3.2.

Given the dimension commonly chosen for the total OSC (amount O_2 per mass of the sample), the drop in OSC could be merely due to the replacement of reducible Ce^{4+} ions by non-reducible Gd^{3+} ions. A calculation of the achieved degree of reduction, i.e. the conversion of Ce^{4+} to Ce^{3+}, however shows that the addition of gadolinium and the associated formation of oxygen vacancies also depresses the reduction of the Ce^{4+} (see Figure 3.6). Both effects, replacement of Ce^{4+} ions by non-reducible Gd^{3+} ions and lower Ce^{4+} conversion thus combine to result in a lower total OSC for the doped materials.

Using the information gathered by the reduction experiments and the known gadolinium content of the samples, one can calculate the number of oxygen vacancies due to (a) the reduction and (b) due to the gadolinium incorporation and thus the total number of vacancies in the host lattice. Table 3.2 lists this information expressed as vacancy "concentrations" for the sample calcined at 550 °C. As can be seen, the total number of vacancies increases with gadolinium addition up to $x_{Gd}=0.3$. This shows that the addition of gadolinium stabilizes the host lattice, and more vacancies can be tolerated. This stabilization effect is believed to be due to the similarity in size between the Gd^{3+} and the Ce^{3+} ions: the lattice strain induced by the creation of larger Ce^{3+} ions is smaller in the doped material because the similarly sized Gd^{3+} ion causes a larger unit-cell volume of the crystal [62] (compared to Ce^{4+}, which has Xeons electron configuration, the reduced Ce^{3+} is larger because one extra atomic orbital is occupied).

While the total number of vacancies is higher, the number of vacancies created by reduction is lower for the doped samples and levels off for $x_{Gd} \geq$

Table 3.2: Comparison of oxygen vacancy concentrations resulting from doping ($V_{O,d}^{\bullet\bullet}$) and reduction ($V_{O,r}^{\bullet\bullet}$), as derived from TG measurements (see Figure 3.6) for the sample calcined at 550 °C

x_{Gd}	concentration of vacancies in mmol/kg		
	$c(V_{O,d}^{\bullet\bullet})$	$c(V_{O,r}^{\bullet\bullet})$	total, $c(V_O^{\bullet\bullet})$
0	0	988	988
0.1	289	769	1058
0.2	575	600	1175
0.25 - run 1	717	519	1236
0.25 - run 2	717	581	1298
0.3	858	575	1433

0.2–0.3 (see Table 3.2). The results for a low gadolinium content fit well with the equilibrium argument above: the presence of vacancies due to the doping with gadolinium hinders the formation of new vacancies and thus hinders the reduction of the material. The positive effect of vacancy stabilization thus is overcompensated by the impaired creation of further vacancies to result in an overall lower OSC for the doped samples. For higher gadolinium fractions close to the incorporation limit though, the vacancy concentration that can be achieved by reduction ($c(V_{O,r}^{\bullet\bullet})$) appears to be fairly independent of the dopant concentration (see Table 3.2). This is surprising, as one would have expected a continuous trend to less and less vacancies created by reduction.

As already mentioned above, the results described above could be merely related to experimental error. An alternative explanation however could be as follows. Regarding the ionic conductivity of the materials, it is observed that with doping, the conductivity increases, but that at higher dopant concentrations, the conductivity is lowered again. In search for an explanation, it is argued that at higher dopant concentrations, the mixed oxide lattice is more stable due to the formation of ordered defect clusters, which results in the formation of "deep vacancy traps" [63], which in turn lowers ionic conductivity.

Considering this in relation to the equilibrium argument of above, a vacancy trap removes the vacancy from taking part in the equilibrium and thus pushes the equilibrium to the right-hand side of Equation 3.2. Vacancy traps therefore should stabilize the reduced state and thus facilitate the creation of new vacancies by reduction. In consequence, it is possible that this mechanism is responsible for the higher than expected vacancy concentration ($V_{O,r}^{\bullet\bullet}$ in Table 3.2) as well as the increase in Ce^{4+} conversion and the levelling-off of the OSC in Figure 3.6. This interpretation corresponds to the dashed grey line in Figure 3.6.

To test the dynamic OEC, further experiments were conducted with alternating H_2 and O_2 pulses. While total OSC is a measure of the maximum amount of oxygen extractable from the sample, the dynamic OEC is a measure of the redox kinetics. The sample is subjected to alternating pulses of H_2 and O_2 and the magnitude of the resulting weight oscillations is the dynamic OEC (see Appendix C for details).

Computational studies have shown that the reduction energy is lowered significantly by doping, facilitating the reduction process [62]. It is thus not surprising that the dynamic OEC increases with the gadolinium content, as shown in Figure 3.7. Interestingly though, the dynamic OEC shows the extremum value for samples with gadolinium content below the maximum solubility of gadolinium in ceria ($x_{Gd}=0.3$, see Figure 3.3). Thus while lower gadolinium contents enhances the reduction kinetics, a high gadolinium concentration thus apparently impairs the reduction kinetics.

Estimating the amount of oxygen needed to reduce the ceria on the surface of the doped sample (about 10 mmol O_2/kg), it becomes clear though that significant bulk reduction also takes place during the above experiments. Further, the results for the dynamic OEC bear some similarity with the ionic conductivity of gadolinium doped ceria, which reaches a maximum value at intermediate dopant concentrations (see Figure 3.7). As mentioned above, one qualitative explanation may be defect clustering as reason for a drop in oxygen mobility at higher dopant levels [63]. Concerning the reduction kinetics, the defect clustering and the resulting lower oxygen mobility would explain the lower values for the dynamic OEC if oxygen diffusion is the rate limiting step.

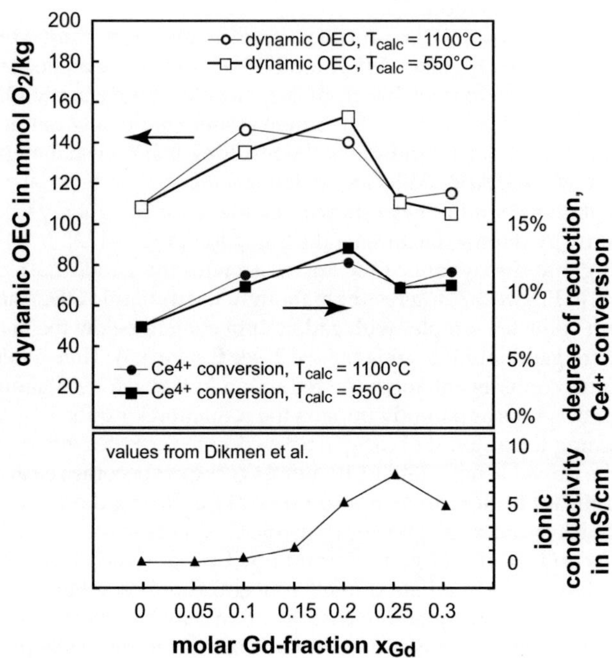

Figure 3.7: Dynamic oxygen exchange capacity and corresponding Ce^{4+} conversion at $800\,^{\circ}C$. For comparison, the ionic conductivity as measured by Dikmen et al. [54] is also shown.

3.1.2. Catalyst stability

One critical aspect of the catalysts examined is their stability under reaction conditions, which was poor in the preliminary study [35]. There are several phenomena that could explain such a deactivation behavior, such as sintering effects or deposition of carbonaceous residues.

Sintering of high surface area ceria has often been reported, especially for reducing conditions [64]. Associated with this sintering of the support, metal encapsulation and metal sintering may take place, resulting in the deactivation of the catalyst. It was found, however, that low surface area ceria is much less affected by high temperature treatment [64] than ceria with higher specific surface area.

The samples under examination here exhibit a rather low specific surface area, thus no extensive sintering of the ceria support is to be expected (see Table 3.1). Sintering of the noble-metal phase, i.e., degradation of the dispersion, can also be a cause of deactivation. However, the dispersion of the catalysts employed is already low (see Table 3.5(b)). Therefore, further particle growth would only have a small effect on the available metal surface.

The catalytic behavior of reducible oxides is characterized by a strong dependence on the reduction temperature prior to the use as catalyst. Essentially, it was determined that upon reduction of the catalyst above a threshold temperature, a strong interaction between the noble metal and the reducible support takes place.

The nature of that interaction and the exact chemical and structural nature of the resulting material is of considerable debate, so the general term "strong metal support interaction" (SMSI), was adopted to describe such phenomena. One common conclusion of studies concerning this subject is that the resulting material is not catalytically active and lacks hydrogen chemisorption capacity. When hydrogen chemisorption is used as means to determine the metal dispersion, the SMSI effect experimentally results in largely underestimated dispersion values. The discrepancy between dispersion determined by hydrogen chemisorption and by other means, e.g. electron microscopy, has often been used as an indication for the "SMSI state" of the catalyst (e.g. [65]).

For noble metal catalysts supported on ceria several mechanisms of SMSI were distinguished [59]. No influence was detected for reduction at temperatures below 500 °C. Upon reduction at temperatures between 500 and 700 °C, an influence on the catalytic properties, but no structural modifications were detected. In consequence, "electronic perturbations" are proposed to be responsible [59]. Upon reduction at 700 °C and higher, structural modifications were revealed.

For rhodium supported on ceria, Bernal et al. showed by high-resolution transmission electron microscopy (HR-TEM) studies the occurrence of metal decoration, i.e. the partial covering of the noble metal by support material [59]. For platinum supported on ceria, alloy formation was observed in addition to metal decoration at high reduction temperatures [59, 64]. In a later study, Penner et al. were able to show that alloy formation also takes place in rhodium/ceria thin-film model systems [66].

Bernal et al. further reported that catalysts supported on terbium doped ceria ($Ce_{0.8}Tb_{0.2}O_{2-\delta}$) and on zirconium doped ceria ($Ce_{0.68}Zr_{0.32}O_2$) were generally only affected by metal decoration at higher temperatures compared to their ceria only analogs. Metal alloying phenomena however were only observed for ceria and terbium doped ceria. For zirconium doped ceria catalysts, the dopant likely stabilizes the ceria against migration onto the surface [59].

For noble metals supported on ceria and on doped ceria used for hydrogen production, such strong metal support interaction phenomena may play an important role not only for the preparation of the catalyst but also during operation, as the reforming of hydrocarbons creates a reducing atmosphere under operating conditions.

Therefore, apart from carbon deposition, the catalysts may also suffer deactivation by metal decoration under the reaction conditions applied. Experiments with conditions similar to the ones employed by Bernal et al., i.e. model studies on catalysts treated in hydrogen, are needed however to allow for comparison of the results.

Model conditions: H_2 chemisorption and HR-TEM results   One indication for the SMSI effect is that after reduction treatment, the capability of the noble metal to dissociatively adsorb hydrogen is progressively suppressed [67]. Generally, a strong reduction of the H_2 chemisorption is observed in most noble metal ceria catalysts for temperatures above 500 °C [59]. A depression of hydrogen adsorption (and consequently of the apparent dispersion) also occurs for the Rh/CGO (x_{Gd}=0.2) catalyst under investigation here, as shown in Figure 3.8. The effect, however, appears to be shifted to higher reduction temperatures; a full depression is observed for the sample reduced at 700 °C. This is in accordance with other studies on doped cerias,

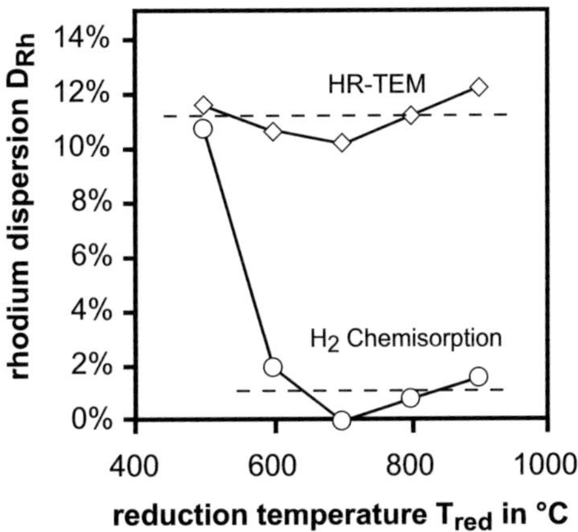

Figure 3.8: Rhodium dispersion obtained by H_2 chemisorption (○) and from HR-TEM (◇). Catalyst: 1 wt.-% Rh CGO x_{Gd}=0.2

which showed that, compared to pure ceria, higher reduction temperatures are needed for Zr- and Tb- doped ceria [59].

It is unexpected that the measured dispersion values for the higher reduction temperatures are increasing. No explanation of this observation is currently available. Measurements of this kind however are always prone to experimental error, it seems therefore more reasonable at this point to consider the values at and above 600 °C to be roughly equal and not quite zero. In other words, the hydrogen adsorption capacity is almost fully suppressed by the SMSI effect for temperatures higher than 600 °C.

As mentioned above, the comparison of dispersion values obtained by chemisorption and by HR-TEM offers some more insight. As shown in Figure 3.8, both methods result in similar values for the dispersion for the catalyst reduced at 500 °C, thus no SMSI is detected. An addition, the similarity between the values obtained by chemisorption and by HR-TEM shows that the adsorption temperature was chosen low enough (see chapter 2.2.2) and that spill-over phenomena influencing the apparent dispersion also can be safely neglected for the purpose of this investigation. At higher temperatures, the dispersion obtained by HR-TEM remains approximately constant. The discrepancy between these values and the ones obtained by H_2 chemisorption thus indicates that the catalyst is subjected to SMSI.

In Figure 3.9, profile view images of catalysts reduced at temperatures between 600 and 900 °C are shown. The sample reduced at 600 °C shows a clean metal surface. Analogous to the interpretation by Bernal et al., reduction at this temperature results in a catalyst in SMSI state, with electronic perturbations being responsible for the reduced apparent dispersion. The catalysts reduced at higher temperatures show a barely discernible covering layer on the noble metal. For these reduction temperatures, metal decoration most likely takes place.

Operating conditions: reaction studies Most studies on ceria supported catalysts consider carbon deposition as a possible deactivation mechanism. As obvious from Figure 3.10, the catalyst's activity degrades rather quickly during steam reforming, as the hydrogen volume fraction in the product gas drops continuously ("first cycle"). To check for the possibility of carbon deposition, the reactor was cooled down (in nitrogen) after the 12-hour run, the feed was switched to air, and the evolution of CO and CO_2 was monitored while the reactor was heated to 870 °C.

As shown in Figure 3.11, almost no CO_2 can be detected, while very small amounts of CO are indicated ("first cycle", Note: the off-set in the signals for the zero point is a systematic error, but was not corrected as it makes the graphs easier to read). The evolution of CO starts at about 500 °C, which is typical for carbon deposits [68].

The total amount of carbon and the resulting catalyst mass loading with carbon, Y_C are shown in Table 3.3 (Note: the off-set mentioned above was now corrected before the integration). After the first run, the total amount of carbon evolved in the first 2 hours of the TPO experiment corresponds to about 0.09 g of carbon per g catalyst. Using the dispersion value of the catalyst (11%), this corresponds to about 850 times the amount of free rhodium on the surface of the catalyst. Carbon deposition therefore might well be the reason for the deactivation in this case.

After the TPO run, the catalyst was cooled under flowing N_2, activated as before in 20 %H_2/N_2 and a second reaction cycle was carried out. The striking difference between this and the first cycle is that the catalyst now gains activity, although on a low level. The following TPO run revealed that compared to the first cycle slightly less carbon was deposited (see Table 3.3).

44

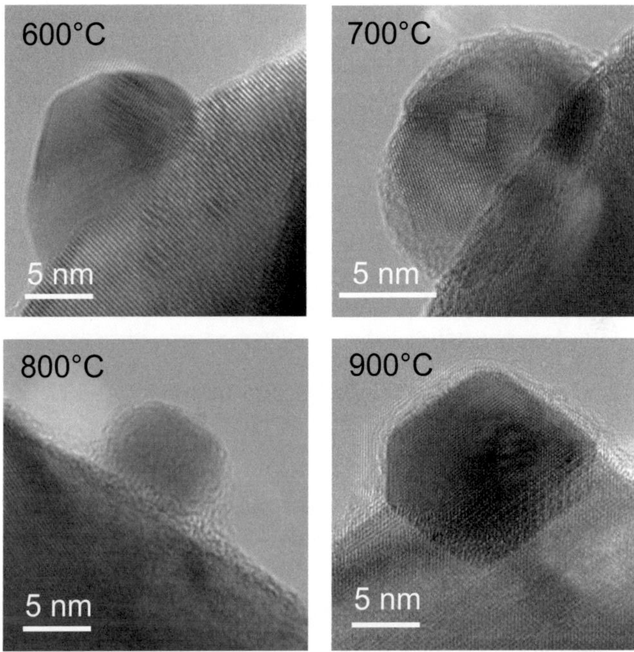

Figure 3.9: HR-TEM images of one Rh particle in profile view in the Rh/CGO x_{Gd}=0.2 samples reduced for 1 hr at the specified temperature.

Table 3.3: Amount of carbon deposited n_C and resulting catalyst mass loading with carbon, Y_C calculated from the TPO experiments for $t \leq$ 120 min.

TPO run after cycle	n_C in mmol	Y_C
1st	2.25	0.090
2nd	1.58	0.063
3rd	-0.02	-0.001

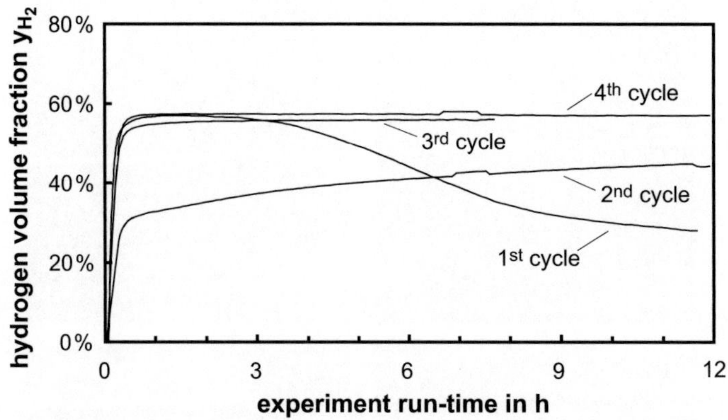

Figure 3.10: Evolution of hydrogen volume fraction in the product gas during steam reforming of natural gas. In between runs, the oxidation run shown in Figure 3.11 and a subsequent reduction was carried out.

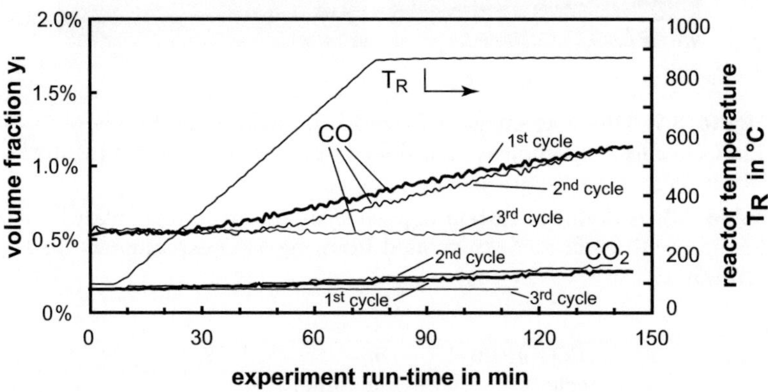

Figure 3.11: CO and CO_2 evolution from the catalyst during TPO, carried out after the reaction cycle shown in Figure 3.10.

During the third reaction cycle, the catalyst's activity now reaches almost the same level as during the first cycle, and no deactivation is detected. In the following TPO run, no carbon is evolved, and during the fourth reaction cycle the catalyst displays stable activity over a period of 12 hours.

Obviously, the redox-cycling of the catalyst must introduce changes in the catalyst, so that no deactivation occurs during the third and the fourth run. Also, no carbon deposition is detected. Thus, one could draw the conclusion that this is the reason for the stability of the catalyst. Bengaard showed for Ni steam reforming catalysts, graphite nucleation takes place at step sites on the Ni crystal, as these are highly active for C-H bond breaking [18]. Assuming that a similar reaction mechanism holds true for Rh catalysts, one possible explanation would be that during the redox-cycling, these highly active sites are removed, e.g. by defect annealing. In consequence, the catalyst is less prone to carbon deposition and therefore more stable. This explanation, however, would imply that the stable catalyst is less active than the fresh one. This is not observed in this case.

In the above experiment, natural gas was used as hydrocarbon feed. The higher hydrocarbons contained are thermally unstable and decompose under reaction conditions even without the presence of a catalyst. With an active catalyst present, however, these decomposition products are converted to CO and H_2 under steam reforming conditions.

Another possible explanation for the behavior described above would be that carbon deposition only takes place on the already deactivated catalyst. In this scheme, the catalyst undergoes a structural change that results in a form that is catalytically inactive for steam reforming. Only then, when the catalyst is inactive, the decomposition products from the higher hydrocarbons cannot react to CO and H_2 and carbon deposition on the catalyst is formed. Carbon deposition would be much more likely for feed containing higher hydrocarbons, because these are much less stable compared to methane. This scheme would also explain why in earlier experiments with pure methane [35] no significant carbon deposition but nevertheless deactivation was observed.

In consequence, the formation of carbon deposits may not be the primary cause of the deactivation, but a result of the already deactivated catalysts. This distinction, however, is difficult in practice as the carbon deposition undoubtedly accelerates the deactivation which couples both processes.

In summary of the observations made under model conditions (see 3.1.2), high temperature reduction causes the catalyst to be in SMSI state, with reduced chemisorption capacity. A catalyst in the SMSI state, a metal

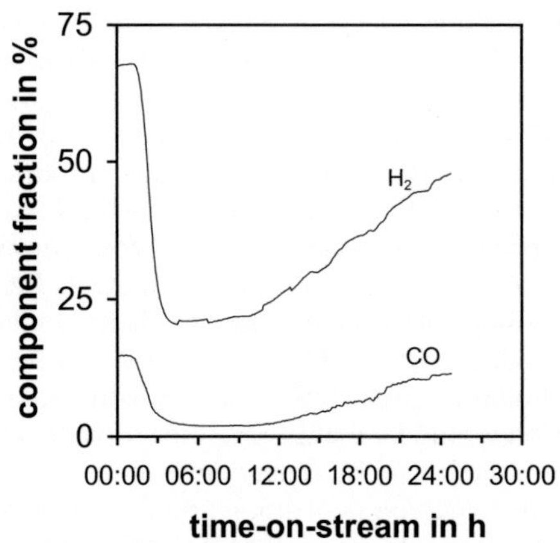

Figure 3.12: Reactivation of the catalyst during reaction conditions. m_{cat}=100 mg, p=1.3 bar, ζ=1:3; feedgas: natural gas H, see Table 2.2

decorated catalyst or a Rh-Ce alloy all could be a catalytically inactive intermediate as argued here. It is therefore proposed that the catalyst upon first use becomes inactive due to the atmosphere under operating conditions. When higher hydrocarbons are contained in the feed, carbon deposition takes place as result of the catalyst's inactivity, while in a case where pure methane is used as feed, no or only very little carbon deposition is detected.

Penner et al. specifically commented on the difference between their observations of Rh-Ce alloys and the results of Bernal et al.[66], who did not observe alloy formation on Rh/ceria catalysts. They pointed out that Bernal's catalysts showed a highly ordered epitaxial relation between the rhodium particle and the support. This results in a very stable arrangement which is probably less prone to alloy formation compared to their catalysts. This catalyst, considering the preparation method, is also very likely not in an ordered state like the samples investigated by Bernal et al. and therefore, by reverse logic, may be very susceptible for deactivation by SMSI.

Continuous redox cycling, however, allows the rhodium particles to achieve a more stable relation to the support. The rhodium particles on the catalyst in this state are much more stable and only to a very small degree subject to metal decoration or alloying. In consequence, the redox treated catalyst displays stable activity under steam-reforming conditions.

This hypothesis would also explain the observed activation during the second cycle in Figure 3.10: if the structural change causing the deactivation is facilitated by the gas phase composition reaction conditions, then the further structural change leading to the stable arrangement should also be taking place under reaction conditions. The reactivation under reaction conditions was indeed observed, see Figure 3.12. Wang [69] and Wisniewski [70] observed similar induction/activation behavior. Thus, redox cycling speeds up the structural change, but is not necessary to facilitate the stabilisation.

Effects of redox cycling: TPR and HR-TEM study If, as argued above, the catalyst undergoes structural changes under reaction conditions and during redox-cycling, these must be visible in a temperature-programmed reduction experiment (TPR). The results of such an experiment are shown in Figure 3.13.

First, the TPR profile of the fresh sample (1st cycle) is considered. Two main features are seen in the TPR profile; the low temperature reduction, typically attributed to the reduction of the noble metal and the high temperature reduction, usually associated with bulk reduction of the support. The onset of the bulk reduction corresponds roughly to the onset of the SMSI state: starting at about 600 °C, the bulk is significantly reduced.

In Figure 3.13, the TPR profiles of subsequent reduction-oxidation cycles are also shown. Between the reduction cycles the catalyst was treated similarly to the above experiment under operating conditions (see 3.1.2), i.e. the catalyst was cooled under flowing nitrogen, oxidized with air (room temperature to 800 °C, 10 K/min) and cooled again before the next reduction run.

Examining the TPR patterns, both onset and extent of the high-temperature reduction seem to be almost uninfluenced by the redox-cycling: the quantitative results in Table 3.4 only vary about 7 %. This confirms that the bulk reduction is, as expected, reversible. Also, the absolute amount of H_2O given off is somewhat lower than what is expected for complete reduction of the material (2300 mmol/kg). Note that this is significantly higher than the values obtained by the TG method (see Figure 3.6), which indicates that the noble metal catalyses the reduction of the support. This effect has been observed also for pure ceria and for ceria-zirconia materials [59, 60]

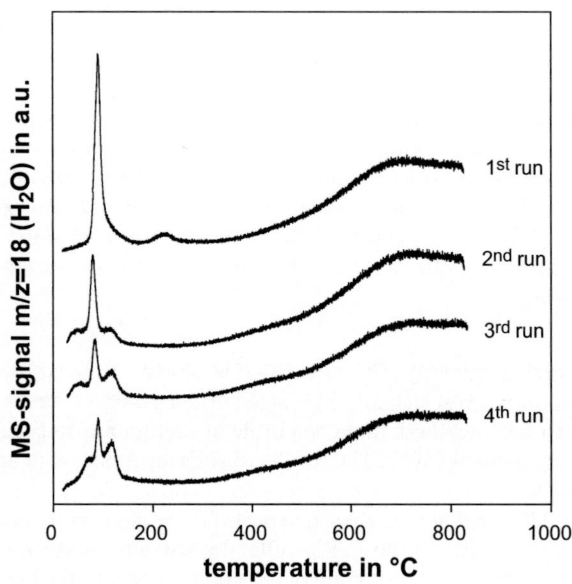

Figure 3.13: Successive TPR experiments on the fresh catalyst.

Table 3.4: Amount of water formed in the low- and high-temperature region during the TPR experiments shown in 3.13. LT-region: RT $\leq$ T $\leq$ 300 °C, HT-region: 300 °C $\leq$ T $\leq$ 800 °C

TPR run	LT-region	HT-region
	n_{H_2O}/m_{cat} in $mmol/kg$	
1st	227	2118
2nd	158	2230
3rd	146	2225
4th	143	2253

Nevertheless, the absence of a clear trend for the four cycles confirms that no changes to the extent of the bulk reduction (and thus the "total oxygen storage capacity") are introduced through the cycling procedure. This result was also to be expected, as the loss of oxygen storage capacity is typically related to support sintering. As this catalyst already has a low specific surface area, no significant sintering takes place, and thus the oxygen storage capacity remains unchanged by the redox-cycling procedure.

For the low temperature behavior, the situation is different. During the first reduction, two distinct peaks, one at about 100 °C, one at 250 °C are visible. The first is likely related to the reduction of the rhodium oxide. This peak coincides with a small peak also visible for the support without noble metal (see Figure 3.5), so some support reduction concurrently takes place. The second, smaller peak at 250 °C in the first reduction cycle must be related to the presence of rhodium, as such a peak is not visible in the TPR profile of the support (see Figure 3.5).

Some considerable changes are introduced during the redox cycles: the first peak significantly shrinks in size, the second peak at 250 °C disappears altogether and at 120 °C, a new shoulder appears. With further redox cycling, the first peak continues to shrink slightly and the new feature at 120 °C increases in size.

For a further evaluation of the TPR spectra, one needs to consider the expected amount of water given off by the rhodium oxide present after synthesis and re-oxidation, respectively. For this catalyst, if only Rh_2O_3 were present, a value of 146 $mmol/kg$ would be expected, if only RhO_2 were present, 194 $mmol/kg$. Looking at Table 3.4, the absolute amount of the water given off in the low-temperature region is slightly larger than these values, which would indicate that some support reduction takes place concurrently.

As during the first cycle more water is released than during the subsequent cycles, one interpretation could be that RhO_2 is present as result of the synthesis, and that after the first reduction and oxidation Rh_2O_3 is formed. However, several other possibilities exist. Given the preparation method of the catalyst, some rhodium may not be located on the surface. Also, the concurrent support reduction complicates the interpretation of the TPR results. The observed changes could therefore also result from a changing fraction of gas-phase accessible rhodium. Further, changes on the support-metal interface would also contribute to the observed phenomena.

In a first attempt to identify changes in the catalyst upon redox cycling, HR-TEM images were taken from a catalyst objected to four times redox cycling at 800 °C. Figure 3.14 shows the resulting images. In these images,

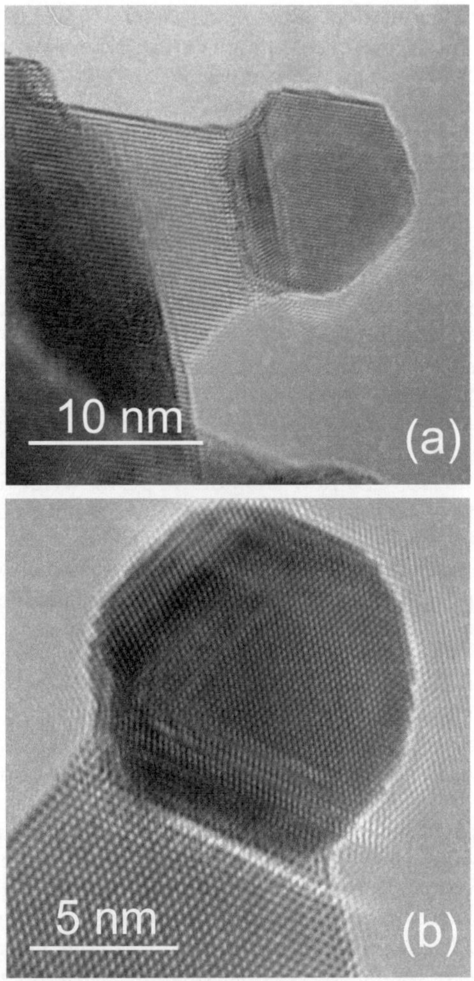

Figure 3.14: TEM image of the catalyst pretreated with 4 redox cycles

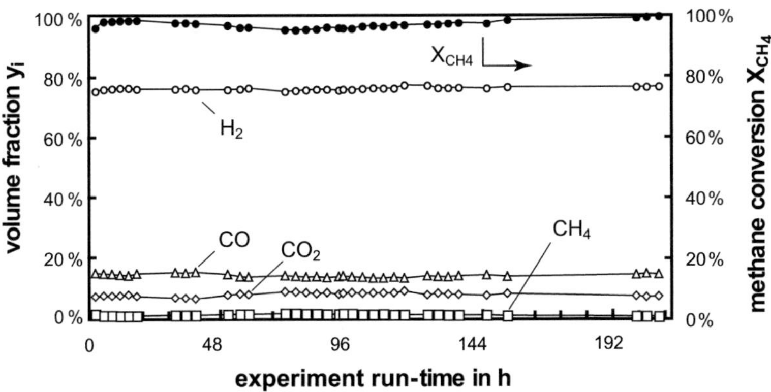

Figure 3.15: Stable performance of the catalyst pretreated with the simplified activation procedure (see text). m_{cat}=100 mg, p=1.3 bar, T_r=800 °C; ζ=1:3; feedgas: natural gas H (composition see Table 2.2)

there is clearly no metal decoration visible, the surface of the rhodium metal appears clean although it was reduced at 800 °C. In Figure 3.14(b), even the step sites on the rhodium particle can be seen very well. This is in contrast to the images shown in Figure 3.9, in which the metal particle appears less defined and possibly covered with a small ceria (or ceria-gadolinium) layer. Note that some periodic structure is seen outside the particle, which is likely an artifact due to the resolution limits of the TEM.

Apart from metal decoration, already discussed above, Bernal et al. commented on two other outstanding nano-structural features of rhodium-ceria catalysts. First, the rhodium particle is grown on the ceria support with one of two possible, defined epitaxial relationships (parallel or 60° rotated). The second feature is that the metal particles seem to be grown on a support pedestal. Both features, a metal particle on top of an exactly fitting support pedestal as well as the epitaxy, are observed for this catalyst after redox cycling: in Figure 3.14(b)) it can clearly be seen that the lattice planes in the support run parallel to the ones in the metal particle, thus showing parallel epitaxy. In comparison, this feature is not seen in the images shown in Figure 3.9, thus it likely is a result of the redox cycling. Equally, the pedestal arrangement is likely a result of the redox cycles, as the metal particles sit flat on the support surface in the images shown in Figure 3.9.

Taking into account the hypothesis outlined above for the reaction studies, these results show that with redox cycling, structural changes occur. The resulting catalyst likely contains more crystalline rhodium species, with the clean rhodium crystals located on a support pedestal.

Long term stability of the redox pretreated catalyst As the redox cycling procedure as outlined above is rather cumbersome to use as a standard activation procedure, a slightly modified procedure was tried and adopted. Instead of both cycling temperature and gas atmosphere, only the gas atmosphere was switched four times between air and 5 vol.-% H_2 in nitrogen (with 5 min of pure nitrogen feed in between), while the temperature was held constant at 800 °C.

In Figure 3.15, the stability of a catalyst treated with this activation procedure for a period of over 190 hours operating time is shown. In addition to a stable methane conversion, it can be seen that the yield of the desired product H_2 and the ratio of CO and CO_2 are also constant, i.e. that the water-gas-shift activity of the catalyst is also stable.

3.1.3. Summary

- The material has no microporosity and a low specific surface area. The primary particles are only weakly agglomerated and no systematic variance of these parameters with the gadolinium content is detected.

- Up to a gadolinium fraction of about $x_{Gd} = 0.3$, the dopant is integrated into the fluorite lattice, and the measured lattice parameters of the doped samples fit published correlations.

- As derived from TPR patterns, the addition of gadolinium ions enhances the reducibility at low temperatures and suppresses the reduction at high temperatures. Further redox measurements in a TG setup at 800 °C showed that the dynamic OEC is enhanced by the addition of low to intermediate amounts of gadolinium, while the total OSC is suppressed.

- Observations made by hydrogen chemisorption experiments and from HR-TEM images for 1 wt% Rh CGO $x_{Gd} = 0.2$ catalysts pretreated under model conditions suggest that high temperature reduction causes the catalyst to be in SMSI state with a strongly reduced chemisorption capacity.

- The freshly prepared catalyst deactivates under typical operating conditions for the steam reforming of natural gas (800 °C, steam to carbon ratio = 3 and natural gas as feed). After three subsequent reduction-reaction-oxidation cycles, however, the catalysts showed stable performance. Results of TPR and HR-TEM experiments suggest that the redox cycled catalyst contains single-crystal rhodium particles, with the clean rhodium crystal located in a defined epitaxial orientation on a support pedestal, while the fresh catalyst contained multi crystal rhodium particles oriented in a less defined fashion on the support. It was however not possible to positively confirm that the above described difference is indeed the cause of the observed improvement in stability.

- Based on the above observations, an activation procedure, consisting of three redox-cycles at 800 °C was tested and will for all following experiments be used. With the developed activation procedure, stable performance of the 1 wt% Rh CGO $x_{Gd} = 0.2$ catalyst for at least 190 hours was demonstrated.

3.2. Activity and sulfur tolerance

To judge the performance of these catalysts, two main reactions have to be considered. First, the steam reforming reaction of hydrocarbons, e.g. for methane

$$CH_4 + H_2O \;\rightleftharpoons\; CO + 3\,H_2 \tag{3.3}$$

is the desired reaction. Further, the product gas composition is always influenced by the water-gas shift equilibrium:

$$CO + H_2O \;\rightleftharpoons\; CO_2 + H_2 \tag{3.4}$$

The conversion of hydrocarbons should be very high for the desired application. Some remaining methane in the fuel-cell feed however might be tolerated, as it is inert in the fuel cell, but may have a beneficial effect on the anode gas burner. The optimum from an overall process perspective might therefore be below full conversion.

The water-gas shift reaction is equilibrium limited under the prevalent conditions, and thus a product gas composition as close as possible to the equilibrium is desired. As another dedicated water-gas shift reactor is usually applied following the reforming reaction, a certain deviation from equilibrium at the reforming reactor outlet can be tolerated.

In addition to the desired reactions above, several other (unwanted) reactions leading to carbon deposits or unsaturated hydrocarbons and also reactions involving sulfur can take place.

3.2.1. Methane and Natural Gas Steam Reforming

Methane steam reforming experiments were carried out for catalysts on supports with a range of gadolinium content, see Figure 3.16(a).

Table 3.5 shows the activation energies derived from Arrhenius plots for the experiments shown in Figure 3.16. From Table 3.5, no clear trend for the activation energy E_A can be derived; up to x_{Gd}=0.2 the activation energy is the same within the experimental error. For the highest gadolinium content, x_{Gd}=0.25, a slightly higher activation energy is measured. The values for this series correlate well with the activation energy obtained by Wei and Iglesia [71], who obtained 108–111 kJ/mol for a rhodium catalyst supported on Al_2O_3. Generally, the four catalysts show similar activity but the fine variation in rhodium dispersion and experimental conditions (e.g.

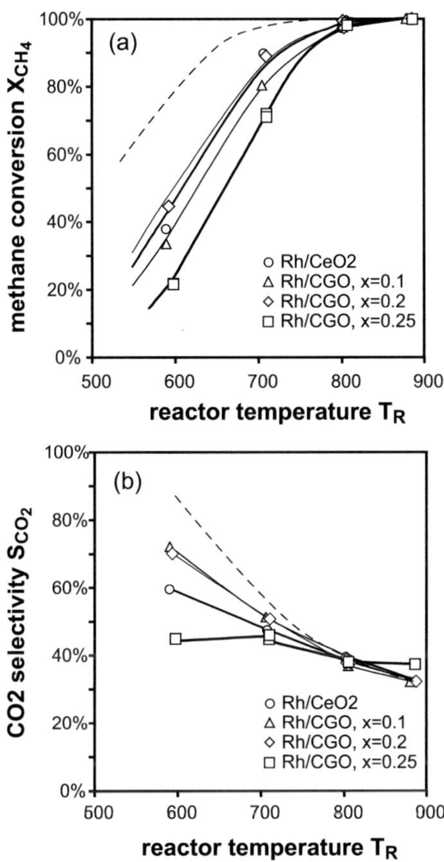

Figure 3.16: (a) conversion and (b) CO_2 selectivity of a series of 1% Rh catalysts with different gadolinium content in the support. $\omega_{Rh} = 1\%$ m_{cat}=100 mg, p=1.3 bar, ζ=1:3; feedgas: 100 vol.-% CH_4. The dashed line represents the expected chemical equilibrium (calculated with the software package ASPEN Engineering Suite, also see Appendix F). Lines for the conversion are calculated using parameters obtained from an Arrhenius plot, lines for the selectivity are simply connecting the data-points.

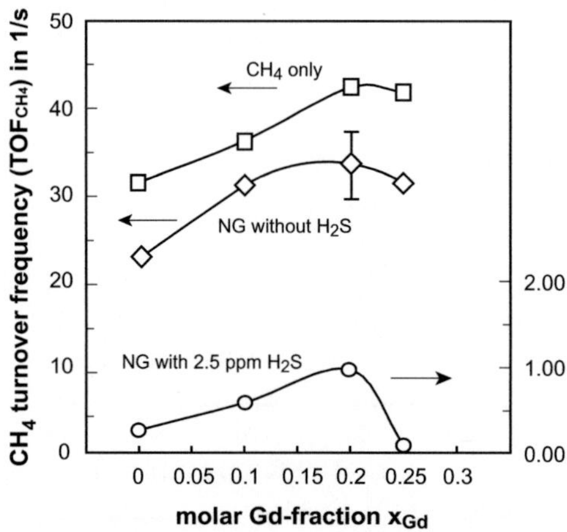

Figure 3.17: Turnover frequencies for several catalysts with varying gadolinium content at 800 °C. (The range given for x_{Gd}=0.2 is derived from the experiments shown in Figure 3.20)

slightly different temperature, catalyst mass, residence time, pressure) blur the comparison. To obtain a better view of the relative performance of the catalysts, an analysis of the turnover frequencies was carried out. As clear from Figure 3.17 (Table 3.5 gives the dispersion measured by the HR-TEM technique), the activity of the catalyst, measured as turnover frequency of methane related to the number of surface rhodium atoms (TOF$_{CH_4}$), there is a variation with the gadolinium content, with the catalyst with x_{Gd}=0.2 being the most active.

This trend is also observed for experiments for which pure methane was substituted for natural gas (composition see Table 2.2). Assuming that the turnover frequency of carbon atoms is constant, a lower turnover frequency of methane is expected in the natural gas case, as the higher hydrocarbons compete with the methane for surface centers. Indeed, a lower TOF for methane is observed for the natural gas experiments. Unfortunately the conversion of the higher hydrocarbons was always complete in the experiments, so experimentally the above explanation cannot be verified.

However mathematically, if a conversion on the order of 99.5% is assumed for both ethane and propane, the carbon turnover frequency (i.e. $\text{TOF}_C = \text{TOF}_{CH_4} + 2\,\text{TOF}_{C_2H_6} + 3\,\text{TOF}_{C_3H_8}$) is close to the measured TOF_{CH_4} in the pure methane case.

The measured turnover frequencies at $600\,^{\circ}\text{C}$ (not shown in Figure 3.17) are similar to the ones obtained by Wei and Iglesia [71]. Extrapolating their result to higher partial pressures assuming a 1st order reaction and further correcting for the lower dispersion in the catalyst examined here (according to the correlation given by Wei and Iglesia in the same paper) gives expected turnover frequencies in the range of 0.6 to $4\,1/\text{s}$. The values measured here (at $600\,^{\circ}\text{C}$) are between 2.3 and $5.8\,1/\text{s}$, which can be considered a rather close agrement. It should be noted that the TOF's calculated here are significantly smaller than the ones calculated for the CO_2 reforming (TOF at $800\,^{\circ}\text{C} \approx 260\,1/\text{s}$, results in [35]). This is probably due to the fact that the TOF's for CO_2 reforming were calculated from an extrapolation to initial activity, due to the deactivating catalyst. The TOF's calculated for the CO_2 reforming thus likely were an overestimation.

The trend of the TOF_{CH_4} with the gadolinium content resembles the trend of the oxygen exchange capacity, OEC, shown in Figure 3.7. As detailed in Section 3.3.1, the oxygen which is part of the support crystal structure participates in the steam reforming reaction. Even more, the enhancement of the OEC by 35–40% corresponds to the enhancement of the TOF. It is thus well possible that the better oxygen exchange capacity - in essence the more loosely bound oyxgen - enhances reactivity. In consequence, this would contrast with the conclusion from Wei and Iglesia that the rate determining step in this case is the methane activation on the rhodium metal [71]. However, Wei and Iglesia used alumina, which is a rather redox-inactive support in their experiments. Thus the above result must not necessarily be in disagreement with their results, it rather suggests that on catalysts containing a redox active support such as ceria-gadolinia, the reaction may proceed differently, with the oxygen transfer from the support being the rate determining step.

One problem of this analysis in general is that the reforming reaction is a structure-sensitive reaction over rhodium: crystallographically different sites contribute differently to the overall observed rate [18, 72]. Therefore, the concept of turnover-frequency needs to be handled with care, because normalizing the reaction rate to the overall number of surface atoms probably constitutes an oversimplification. Another problem is that many measurements have to be done to calculate the TOF, so the measure-

Table 3.5: Activation energy (a) and dispersion (b) of several rhodium catalysts for methane reforming.

(a)

x_{Gd}	E_A in kJ/mol
0	89
0.1	93
0.2	83
0.25	115

(b)

x_{Gd}	D_{Rh} in %
0	15.6
0.1	11.5
0.2	11.1
0.25	10.6

ment errors add up to some considerable uncertainty. Appendix G details a calculation which yields an uncertainty of $\approx \pm 19\,1/\mathrm{s}$ on the calculated TOF_{CH_4}. This uncertainty is higher than the differences seen in Figure 3.17, allowing for no conclusion about the relative ranking of the catalysts. On the other hand, some of the deviations are systematic and thus the same for all catalysts. Taking this into account, the uncertainty (in regards of comparing the catalysts) reduces to about $\pm 8.2\,1/\mathrm{s}$, shown as error bar in Figure 3.17. Considering this uncertainty it appears as if the sample with a gadolinium content of $x_{Gd}=0.2$ is the most active sample.

Apart from the activity for the steam reforming reaction, the selectivity towards hydrogen, in other words the extent to which the water-gas-shift equilibrium (Equation 3.4) is reached, is important. Experimentally, the selectivity towards CO_2 is an easier measure, and it expresses the same. If the selectivity towards CO_2 is high, there is more hydrogen formed. Figure 3.16(b) shows the measured CO_2 selectivity and the chemical equilibrium calculated with the measured methane conversion.

The measured selectivities for all catalysts are roughly similar to the ones expected in chemical equilibrium for 800 and 900 °C. At lower temperatures though, the measured values deviate substantially from the calculated equilibrium, indicating that the kinetics of the shift reaction are not fast enough to reach equilibrium. (Note that the method of calculating the chemical equilibrium is important - see Figure F.3 in Appendix F.)

In Figure 3.17, the activity after addition of hydrogen sulfide to the feed is also shown. The activity is significantly lowered, but a small remaining activity can still be observed. Interestingly, the activity ranking as discussed above between the catalysts with different gadolinium content remains.

Further, the drop in activity depends on the gadolinium content. For the activity of the pure ceria with sulfur is $\approx 1\,\%$ of the one without sulfur, for $x_{Gd}=0.2$, the remaining activity is still $\approx 2.5\,\%$.

3.2.2. Dynamics and reversibility of the sulfur poisoning

As seen in Figure 3.18, the changes upon sulfur addition to the feed proceed in several stages. Directly after addition of H_2S a detrimental effect is visible on the water-gas shift equilibrium (Equation 3.4); the product distribution is shifted from CO_2 to CO. During this period, almost no change in methane content is detected. Further, no H_2S is detected in the product gas, most likely because it is adsorbed on the catalyst.

Once H_2S breaks through (see bottom part of Figure 3.18), the methane content rises sharply, i.e. the methane conversion (Equation 3.3) drops. This indicates that the adsorbed sulfur has deactivated the catalyst. The lower conversion now causes both the CO and the CO_2 content in the product gas to decrease. Due to the dropping conversion, the H_2 content in the product also falls.

The drop in methane conversion however does not proceed towards zero conversion, but reaches a new steady state. Also, the amount of H_2S in the product gas corresponds closely to the amount of H_2S added to the feed; the H_2S balance error is $0.7\,\%$ (Note: this value is exceptionally low, but an acceptable error was achieved regularly, see also Table 3.6). Calculating the integral amount of H_2S that entered the reactor minus the amount measured at the outlet reveals that a surface coverage of rhodium $\theta_{Rh,S} \approx 23\%$ is reached. This means that, assuming a stoichiometry of 1 sulfur atom per rhodium atom and the dispersion shown in Table 3.5, approximately one in four surface rhodium atoms carries an adsorbed sulfur atom.

This is in line with the common view on sulfur poisoning as competitive adsorption, with the sulfur atom as poison adsorbing significantly stronger than the reactants under reaction conditions [21]. As a result, adsorption equilibrium (see Equation 3.5) lies far on the side of the adsorbed sulfur atom, "Rh-S":

$$H_2S + Rh^* \; \rightleftharpoons \; Rh - S + H_2 \qquad (3.5)$$

In Figure 3.18, once the H_2S addition to the feed is stopped, the methane conversion recovers quickly. Consequently, the H_2 content in the gas also increases. As expected, the rate of recovery is slowing with time, but full

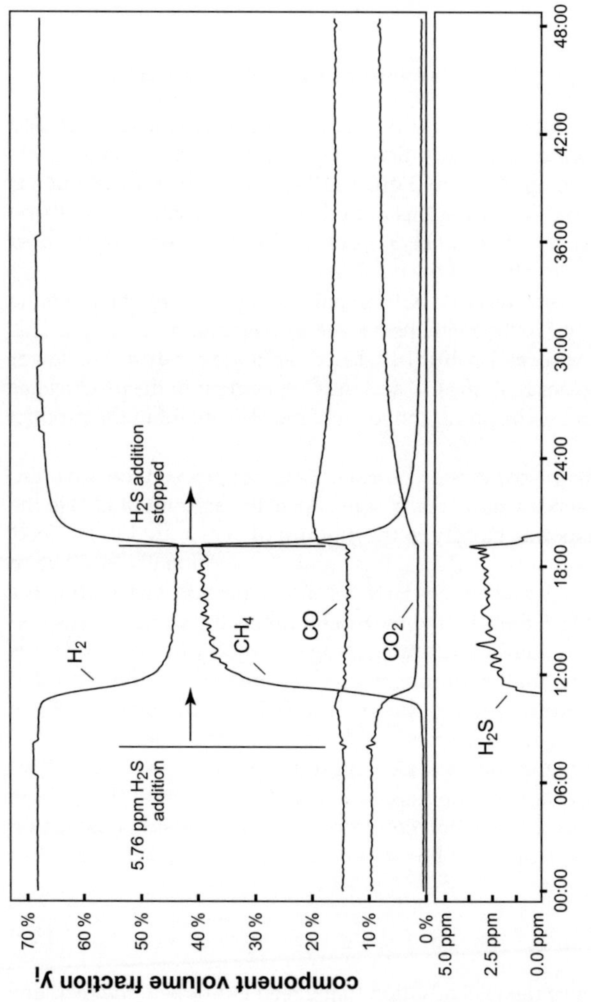

Figure 3.18: Development of product gas composition upon addition of 5.66 ppm H$_2$S (equivalent to 8 mg/m^3 sulfur) to the feed and when addition was stopped at the indicated time, respectively. Reaction conditions: T_R=800 °C, $p_{R,out}$=1.3 bar, τ_{mod}=0.07 (g·s)/ml, $\zeta = 3$, m_{cat}=300 mg, feedgas composition: see Table 2.2.

recovery to the same level of before H_2S addition is reached. The recovery of the shift-reaction, takes significantly longer, as evidenced by the long time it takes for the CO and CO_2 contents to reach their original values.

Interestingly, H_2S is detected in the product gas for a short period only after the addition was stopped. The amount of H_2S measured at the outlet corresponds to only 18% of the adsorbed H_2S, which would mean that a sulfur coverage of $\theta_{Rh,S} \approx 19\%$ is maintained. However, considering a detection limit of H_2S of 1ppm (determined by the Micro-GC used), it would take between 4 hours (1 ppm·90%) and 35 hours (1 ppm·10%) for the rest of the sulfur to desorb. This is the same time scale as the recovery of the shift reaction, so there is indirect indication that all the sulfur is desorbing from the catalyst. Also, elemental analysis by EDX did not reveal any sulfur on a used catalyst sample. This is however not a very strong proof due to the high vacuum conditions in the transition electron microscope.

3.2.3. Long-term stability under sulfur loaded feed

In the experiment whose results are shown in Figure 3.18 and described above, the catalyst was only exposed to sulfur containing feed for a limited time-span. In Figure 3.19, the gas composition and the methane conversion are shown when tetrahydrothiophene (THT), a typical odorant used in Europe is added to the feed (a sulfur loading of Y_S=15 mg/m³, which is typical for odorized natural gas, was used). A constant conversion is maintained for longer than 100 h.

Using the dispersion value shown in Table 3.5(b), it can be determined that enough sulfur was introduced into the reactor within the first 8.5 hours to entirely cover the metal surface area with adsorbed sulfur (assuming an adsorption stoichiometry of S:Rh = 1). The results thus confirm that the catalyst shows sulfur-tolerance, since not only a steady state is reached, but also the reforming activity is maintained much beyond the time of theoretical complete surface coverage.

Further, in Figure 3.20, a much longer time-span is covered. Between the experiments shown in Section 3.4, a standard condition was set for at least 12 hours, resulting in the plot shown. Despite the fact that large amounts of sulfur, resulting in almost no activity, were added to the feed numerous times(see Section 3.4), the catalyst regains its previous activity after some time needed for recovery.

During the time-span shown above, the catalyst was exposed to a N_2/steam mixture several times because the measured pressure drop across the cat-

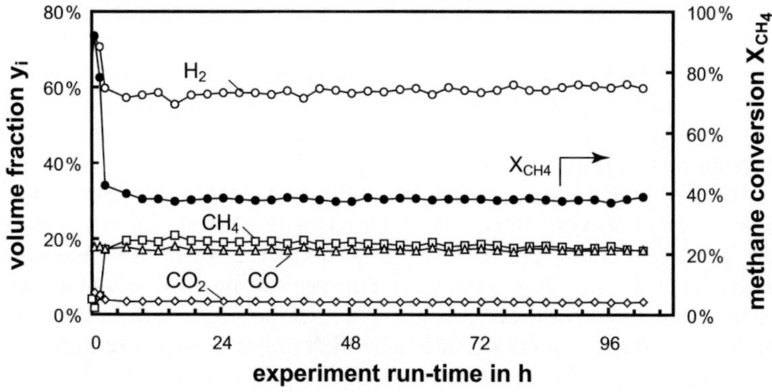

Figure 3.19: Gas composition at reactor outlet and methane conversion for a natural gas containing $15\,\mathrm{mg/m^3}$ sulfur, added as THT. Stable conversion for more than 100h. T_R=800 °C, $p_{R,out}$=1.3 bar, τ_{mod}=0.13 (g·s)/ml, $\zeta = 3$, feedgas composition: see Table 2.2.

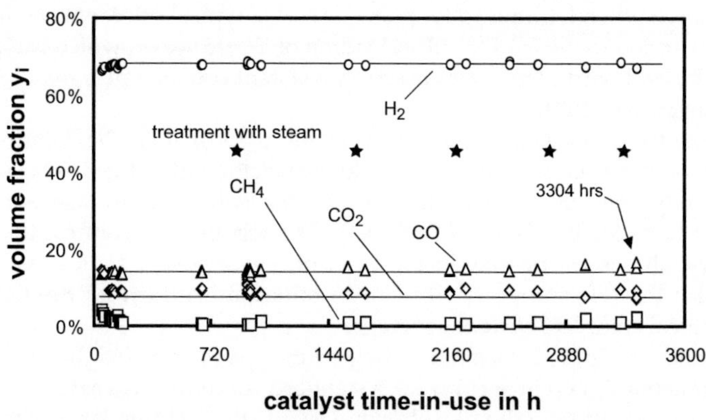

Figure 3.20: Gas composition at reactor outlet and methane conversion over the entire duration of the experiments. T_R=800 °C, $p_{R,out}$=1.3 bar, τ_{mod}=0.13 (g·s)/ml, $\zeta = 3$, feedgas composition see Table 2.2.

alyst bed indicated some carbon deposition (in Figure 3.20 indicated by a star). Among the conditions leading to the carbon deposition were mainly issues related to the operation of the reactor, such as failure of the water pressure or a clogged water nozzle, resulting in only natural gas and no water being fed and thus inevitably leading to carbon deposition. It is noteworthy though that during the operation at a low steam to carbon ratio ζ or with large amounts of sulfur added to the feed, a steady increase in pressure drop, indicating carbon deposition, was also observed.

Although the necessity for steaming the catalyst several times during the 3300 hours prevents the conclusion that the catalyst was under operation under severe conditions all this time without the need for regeneration, it is clear that the catalyst under investigation is very robust.

3.2.4. Summary

- measurements on a series of 1 wt.-% Rh CGO catalysts with different gadolinium content show that there is no clear trend for the the activation energy with the gadolinium content; the measured values 83–115 kJ/mol are close to values found in literature.

- the water gas shift reaction is in equilibrium for the chosen conditions.

- an analysis of the turnover frequencies for the series gives similar values for all catalysts. Although the results suggest that the gadolinium has a positive influence on the methane reforming reaction (the catalyst with $x_{Gd}=0.2$ displays the highest TOF), the uncertainty in the measurement does not allow for definite conclusions.

- The development of the product gas composition upon addition of sulfur to the feed revealed that the water-gas shift reaction is immediately affected. After some time, the methane conversion is also strongly reduced, while H_2S is observed in the effluent. Once the sulfur addition is stopped, the methane conversion recovers relatively quickly, while the shift-reaction requires longer to return to the original level. The effect of sulfur thus is reversible.

- longer experiments show that the catalyst maintains activity despite continuous addition of sulfur containing compounds to the feed.

3.3. Transient Pulse Analysis

To gain more insight into how the catalysts interact with the reactants, transient pulse analysis was carried out. The catalyst is mounted in a small reactor, heated and subjected to a flow of inert carrier gas. Then reactants are pulsed into the inert flow by means of a syringe, while the effluent of the reactor is monitored by a mass-spectrometer. Before these experiments, similar pulse experiments were carried out with an empty reactor, to test for thermal conversion or impurities in the gases used (see Appendix E).

3.3.1. Interaction with methane - coking resistance

All metal catalysts are prone to deactivation by carbon deposition. Concerning the mechanism of deactivation by carbon deposits during steam reforming, it was proposed that hydrocarbons decompose on the metal to adsorbed carbon species, which react with previously adsorbed oxygen to the product CO. If the supply of oxygen on the metal surface is too low, several adsorbed carbon species can form polymers, which, depending on their size, become very stable and cannot react with oxygen to CO. These polymers then block the metal surface and deactivate the catalyst.

The propensity of a catalyst to deactivation by coking, therefore, is the result of a balance between carbon forming reactions (such as methane dissociation) and carbon consuming reactions (such as CO formation or "gasification") [1]. Typically, a large oxygen excess is used to ensure sufficient oxygen supply. For example, in steam reforming a steam to carbon ratio in the feed $\zeta = 3$ is typical [8]. Further, the support and/or promoters are chosen such that the adsorption of water on the catalyst is facilitated, e.g. by using magnesium aluminum oxides ($MgAl_2O_4$) as support or by adding alkali metals such as K or Ca as promoters [1, 73].

As extension to this approach, Roh et al. used a nickel catalyst supported on a support with oxygen storage capacity at the typical operating temperature of 800 °C (Ce-Zr-O), and proposed that lattice oxygen may take part in the reaction, oxidizing adsorbed carbon [26]. The polymerization of carbon species thus becomes less likely, and the reaction could be run with a lower excess of water (a lower steam-to-carbon ratio) in the feed without loss of catalyst stability.

Figure 3.21 shows the mass-spectrometer traces for several methane pulses, for the indicated catalysts. Before the methane pulses, the catalyst was reduced using hydrogen. Both the catalyst supported on ceria and on

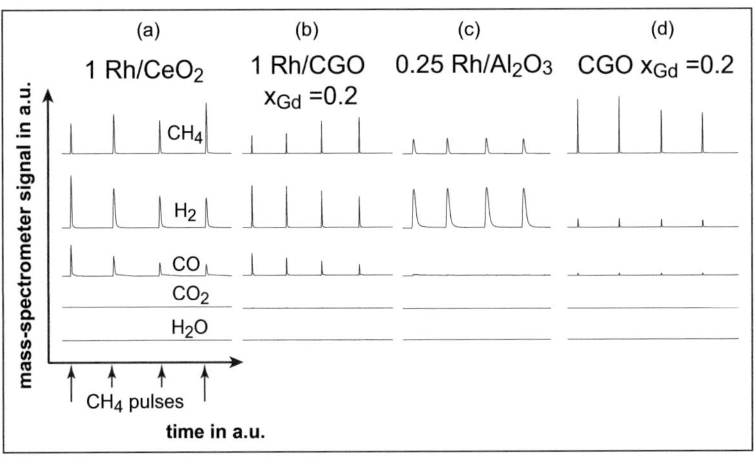

Figure 3.21: Mass-spectrometer traces (measured signal intensity plotted against time) at the outlet of the reactor after injection of CH$_4$.

doped ceria show reforming activity: both hydrogen and carbon monoxide are measured. In contrast, for the catalyst supported on alumina, only hydrogen is measured.

Upon contact with rhodium, methane disassociatively adsorbes. The hydrogen atoms subsequently combine and desorb as molecular hydrogen, while the carbon remains adsorbed on the rhodium surface.

$$Rh* + CH_4 \leftrightharpoons Rh - C + H_2 \tag{3.6}$$

For the alumina catalyst, which is not reducible under the prevalent conditions, no further reaction is possible. For the ceria catalysts, the situation is different, as the catalyst support is reducible under the prevalent conditions (see Sections 3.1.1 and 3.1.1).

In accordance with the interpretation put forward by Roh et al. [26], at least some of the adsorbed carbon can react with the oxygen from the support, forming carbon monoxide, which desorbs and is measured at the exit of the reactor, and leaving an oxygen vacancy in the support.

$$Ce_{0.8}Gd_{0.2}O_{1.9} + Rh - C \leftrightharpoons Ce_{0.8}Gd_{0.2}O_{1.9-\delta} + Rh* + CO \tag{3.7}$$

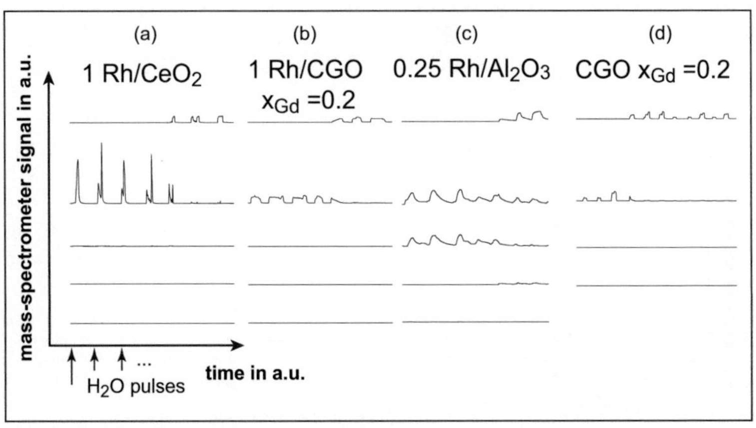

Figure 3.22: Mass-spectrometer traces at the outlet of the reactor after injection of H_2O subsequent to the CH_4 pulses shown in Figure 3.21.

or in Kröger-Vink notation:

$$O_O + Rh - C \leftrightharpoons V_O^{\bullet\bullet} + Rh * + CO \tag{3.8}$$

To check whether carbon is left on the surface on the catalyst, water was pulsed on the catalyst (see Figure 3.22). As obvious from the mass-spectrometer traces, neither CO nor CO_2 are formed for the catalysts on reducible support, indicating that no carbon was present on the surface of the catalyst. For the catalyst supported on alumina, on the other hand, CO is detected, which indicates that, as stated above, the methane decomposition left carbon adsorbed on the surface of the catalyst. Along with the CO, hydrogen is observed, as expected.

$$Rh - C + H_2O \leftrightharpoons Rh * + CO + H_2 \tag{3.9}$$

For the catalyst supported on ceria and doped ceria, liberation of hydrogen is measured nevertheless, indicating that the water oxidizes the support, filling the oxygen vacancy left by the carbon oxidation, Reaction 3.10/3.11.

$$Ce_{0.8}Gd_{0.2}O_{1.9-\delta} + H_2O \leftrightharpoons Ce_{0.8}Gd_{0.2}O_{1.9} + H_2 \tag{3.10}$$

or in Kröger-Vink notation:

$$V_O^{\bullet\bullet} + H_2O \leftrightharpoons O_O + H_2 \tag{3.11}$$

The unsteady shape of the mass-spectrometer pulses seen in Figure 3.22 are likely due to the fact that liquid water was pulsed directly with a syringe in the flow upsteam the hot reactor. This causes evaporation to take place inside the syringe needle, giving rise to pressure variations that then show in the mass spectrometer traces. While more smooth peak shapes would likely have been possible with a more refined experimental technique, the experiments shown above still display the principle of the underlying mechanism.

In Figures 3.21 and 3.22, the same experiments as above, but using the support material only, are also shown. Interestingly, the catalyst still shows some activity, indicating that the support alone can facilitate the reaction. From the peak heights, however, it is clear that the support is much less active than the metal supported on the same material. This is an important result, as it shows that methane mainly interacts with the metal. This result was expected and is in line with the common thinking about the steam reforming reaction mechanism on transition metals.

The pulse experiments reported above confirmed that the ceria and ceria-gadolinia supports can take part in the steam reforming reaction network, supplying oxygen to oxidize carbon on the catalyst's surface. This is in contrast to non-reducible supports such as Al_2O_3. Figure 3.23 shows a simple schematic.

To check whether this indeed translates in a higher resistance towards deactivation by carbon deposition, the catalyst supported on doped ceria (x_{Gd}=0.2, 1 wt.-% Rh) was used for steam reforming under harsh conditions, using a steam to carbon ratio ζ=1.5. As can be concluded from Figure 3.24, the catalyst exhibits rather stable behavior, with only very slight deactivation visible during the 100 h of the experiment. The redox properties of the support obviously have a positive impact on the stability of the catalyst towards deactivation by carbon deposits.

3.3.2. Interaction with sulfur containing compounds

Interaction with H_2S only The interaction of the catalyst with H_2S at $800\,°C$ can be deduced from Figure 3.25(a). As expected, the H_2S dissociates into hydrogen and sulfur, the hydrogen leaves as atomic hydrogen via the

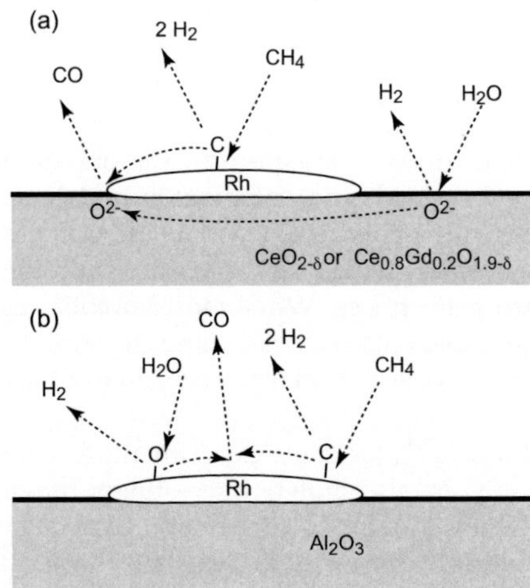

Figure 3.23: Schematic of the reactions (a) on a reducible support involving lattice oxygen, and (b) on a non-reducible support.

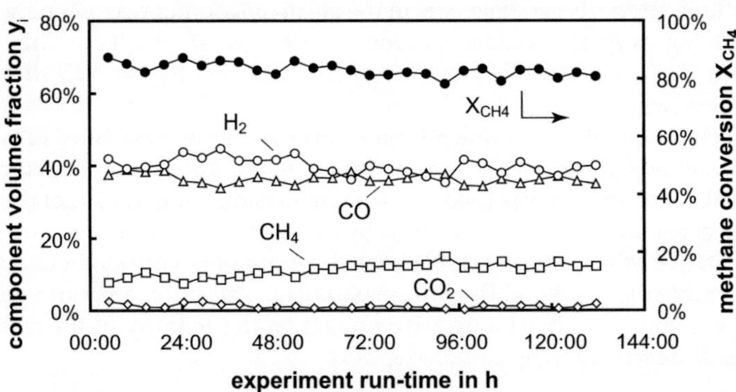

Figure 3.24: Only slight deactivation is observed for the catalyst operated at a steam-to-carbon ratio ζ of 1.5. T_R=850 °C, $p_{R,out}$=1.3 bar, τ_{mod}=0.015 (g·s)/ml, feedgas composition: y_{CH_4}=78 vol.-%, y_{N_2}=22 vol.-%

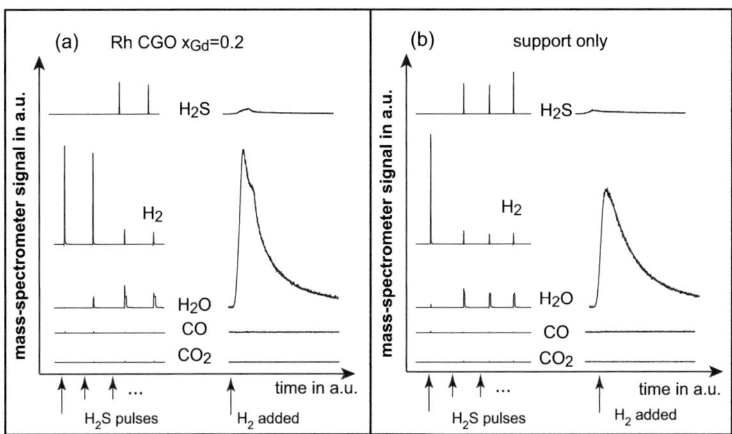

Figure 3.25: H_2S pulse experiments with the CGO (x_{Gd}=0.2) material: (a) 1wt.-% Rh CGO catalyst, (b) support only.

gas phase. No hydrogen sulfide is detected in the reactor outlet, thus the sulfur likely stays adsorbed on the catalyst. In a subsequent flushing of the reactor with hydrogen (see Figure 3.25(b)) hydrogen sulfide leaves the reactor. This is a clear indication that sulfur indeed stayed adsorbed on the catalyst surface and reacted back to hydrogen sulfide as soon as the hydrogen partial pressure in the gas phase was high. At the same time, the support is reduced, as evident from the large amount of water formed. The above findings thus can be explained by the sulfur adsorption equilibrium on rhodium metal (Equation 3.5) and by the reduction of the support by hydrogen, forming oxygen vacancies (Equation 3.2).

It is interesting to note that one H_2S injection corresponds to 41.5 μmol sulfur, while the loaded catalyst only contains approximately 9.7 μmol of rhodium, of which about 11.5% are at the surface. This implies that the rhodium surface is likely full of sulfur after the first pulse. Furthermore, there is too little rhodium surface available to adsorb all sulfur; and as there is no hydrogen sulfide detected at the outlet, the sulfur must be adsorbed on the support.

Looking at Figure 3.25(b), the support alone shows the same behavior as the catalyst at and after the second pulse, indicating that indeed rhodium is not needed for the interaction. This corresponds to a reaction where hydrogen sulfide reacts directly with the support forming a new vacancy, which is subsequently filled with sulfur forming Ce-S bonds on the support surface and water. This is equivalent to the support acting as a high-temperature sulfur adsorbent, which would correspond to a reaction equation like Equation 3.2, only with hydrogen sulfide instead of water and with sulfur occupying an oxygen lattice position, indicated by "S_O":

$$Ce_{Ce} + 2O_O + \delta/2H_2S \rightleftharpoons Ce'_{Ce} + (2 - \delta)\,O_O + \delta S_O + \delta/2H_2O \quad (3.12)$$

Karjalainen et al. showed in thermodynamic equilibrium calculations that ceria has the tendency to form $Ce(SO_2)_3$ and Ce_2O_2S species in reducing atmosphere. Further, this behavior is not altered in the presence of noble-metals; thermodynamically the interaction of ceria with sulfur is more favorable than reactions between the noble-metal and sulfur [74]. Their result ties in nicely with the above observation, except that it seems that the interaction with rhodium is taking place in parallel and that the interaction of sulfur with the support is not very strong. The sulfur is driven off as H_2S once enough hydrogen is present in the gas phase.

Summarizing the above experiments, it can be concluded that hydrogen sulfide reacts with both the rhodium metal and the support. Further it can be concluded that the reaction with the support is reversible; whether the reaction with rhodium is reversible cannot be determined from these experiments. However it is clear from the regeneration of the steam reforming activity (see Section 3.2.2 and Figure 3.18) that the interaction with rhodium also is reversible.

Figure 3.26 shows the same experiments, but for the undoped ceria as support. Essentially, the same effects as for the doped ceria are seen for pure ceria. There is however a difference: the pure ceria adsorbes much more sulfur, over both the catalyst and the support only. This is consistent with the explanation above and with the results in Section 3.1.1 (Figure 3.6) that showed that undoped ceria offers a higher oxygen storage capacity than doped ceria, which means that more oxygen vacancies can be created. This in turn means that undoped ceria can adsorb more sulfur.

As second important result, almost no H_2S is found in the gas-phase once hydrogen is flushed over the undoped support material (Figure 3.26(b)). This shows that the sulfur is bound more strongly to the undoped support -

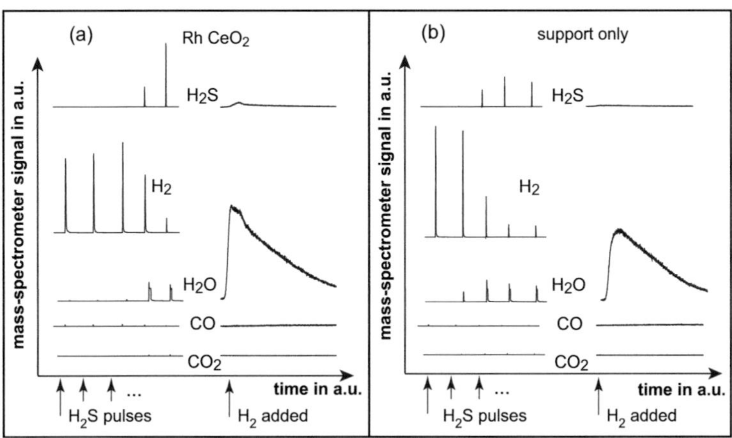

Figure 3.26: H_2S pulse experiments with ceria: (a) 1wt.-% Rh CeO_2 catalyst, (b) support only.

the sulfur adsorption is less reversible for this catalyst. This ties in with the results for the oxygen exchange capacity, shown in Section 3.1.1 (Figure 3.7). Oxygen is more easily exchanged between gas-phase and support in the case of the doped ceria, and undoped ceria has a stronger bond between oxygen and ceria. This corresponds to the behavior towards sulfur: the undoped support apparently binds sulfur stronger than the doped one, resulting in less reversibility.

In the experiments above (see Section 3.2.2) it was shown that sulfur addition inhibits the water-gas shift reaction strongly. Gorte and Zhao [75] reviewed the application of ceria supported noble metals as catalysts for the water-gas shift reaction in fuel cell applications. Essentially, two types of reaction mechanisms were proposed; it is still unclear which is more likely. However, one of the mechanisms, the one favoured by Gorte and Zhao, is a redox mechanism, in which CO adsorbs on the noble metal and is oxidized using crystal lattice oxygen. The support is reoxidized by the water, releasing hydrogen. In addition, Gorte and Zhao observed themselves that the noble metal - ceria interface is very important, both ceria and the noble metal (in their case palladium) showed orders of magnitude less activtiy than the ceria supported metal. These observations tie in very closely with the results here: if sulfur interacts with the support and occupies oxygen lattice positions, the oxidation step to CO_2 becomes inhibited, and the observed water gas shift activity is reduced. Once the sulfur addition is stopped, the sulfur on the support is slowly replaced by oxygen, as evidenced by the recovery of the water gas shift activity (see Section 3.2.2 and Figure 3.18).

Interaction with other sulfur compounds   Figure 3.27 shows the resulting mass spectrometer traces when the indicated sulfurous compound is pulsed on the catalyst. As predicted by analogy from literature results [22], all compounds decompose under steam reforming conditions, forming hydrogen, CO and sulfur, which adsorbs on the catalyst similarly to the hydrogen sulfide case, discussed above. As some of the compounds have water as impurity, CO_2 also is among the reaction products.

Note that the compounds were pulsed onto the catalyst without intermediate regeneration by steam, so the reforming activity of the catalyst becomes deactivated, resulting in unreacted hydrocarbon (methyl fragments) in the effluent.

3.3.3. Methane reforming and hydrogen sulfide addition

To determine the behavior of the catalyst once the steam reforming reaction is taking place on the catalyst in the presence of sulfur, several more pulse experiments are carried out, now with methane, water and hydrogen sulfide pulses in sequence.

Figure 3.28 shows the results of the following series of pulse experiments. First the catalyst was reduced and brought to 800 °C under argon flow. Then, a methane pulse and subsequently a water pulse were injected into the inert flow. The resulting mass-spectrometer traces are shown to the left and are similar to the results shown in Figures 3.21 and 3.22 - methane reacts 'dry' with the support to form CO, and water re-oxidises the support, releasing hydrogen. Then, 1 pulse H_2S was injected over the catalyst. Subsequently, a methane and then a water pulse are injected, the resulting traces are seen in the second column of Figure 3.28. As expected, the activity of the catalyst is severely reduced, but qualitatively methane still reacts with the support to form CO and water re-oxidises the support releasing hydrogen. As

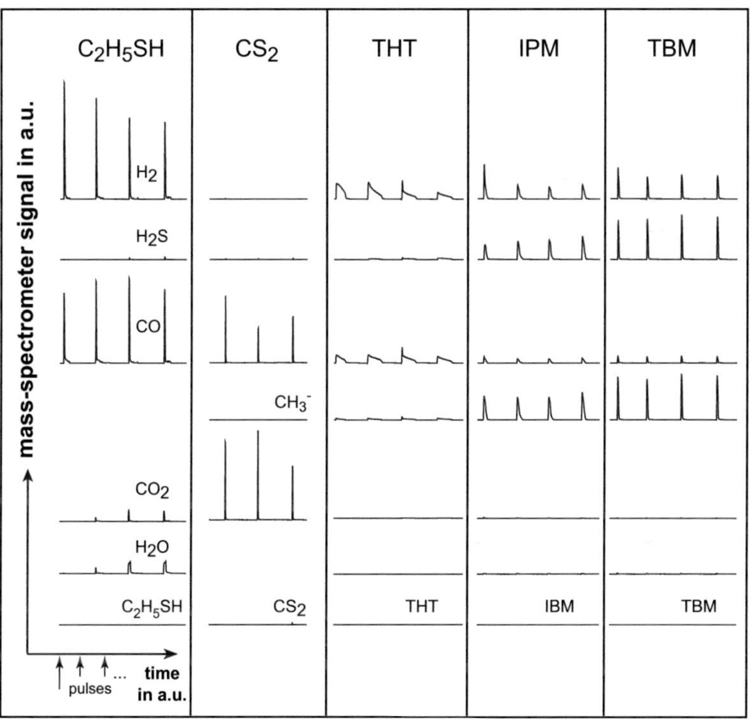

Figure 3.27: Decomposition of various sulfur compounds on a 1wt.-% Rh CGO (x_{Gd}=0.2) catalyst. IPM: iso-propylmercaptan, C_3H_7SH; THT: tetrahydrotiophene, C_4H_8S; TBM: tert-butylmercaptan, C_4H_9SH.

discussed above, one H_2S pulse is more than enough to cover the rhodium surface with sulfur. This result thus confirms that the rhodium retains reactivity towards methane even though it was exposed to sulfur. In contrast to the sulfur free case, some H_2O and CO_2 is seen, but this effect is likely a result of the injected H_2S containing water as impurity, not a result of modified catalyst properties.

After a total of 5 H_2S - CH_4 - H_2O cycles, one more change can be identified: both during the methane and the water pulse some H_2S is released from the catalyst. This indicates that water (or the subsequent reaction product H_2) reacts with adsorbed sulfur on the catalyst to form H_2S, which leaves the reactor. As the catalyst still shows reactivity towards methane, and still forms CO and H_2, this result is another clear indication of sulfur tolerance. Further, this behavior does not change during even more H_2S - CH_4 - H_2O cycles; after a total of 35 cycles the resulting traces are essentially the same as the ones after 5 cycles.

To check for the capability of the catalyst to regenerate once sulfur is not added to the feed anymore, 8 methane - water cycles without H_2S were applied. As can be seen, the catalyst does regenerate its activity to some degree, however not to its original activity. Given the results shown in Figure 3.18 though, it is possible that the 8 cycles are not enough sulfur free gas, and that after more cycles a stronger effect would have occurred.

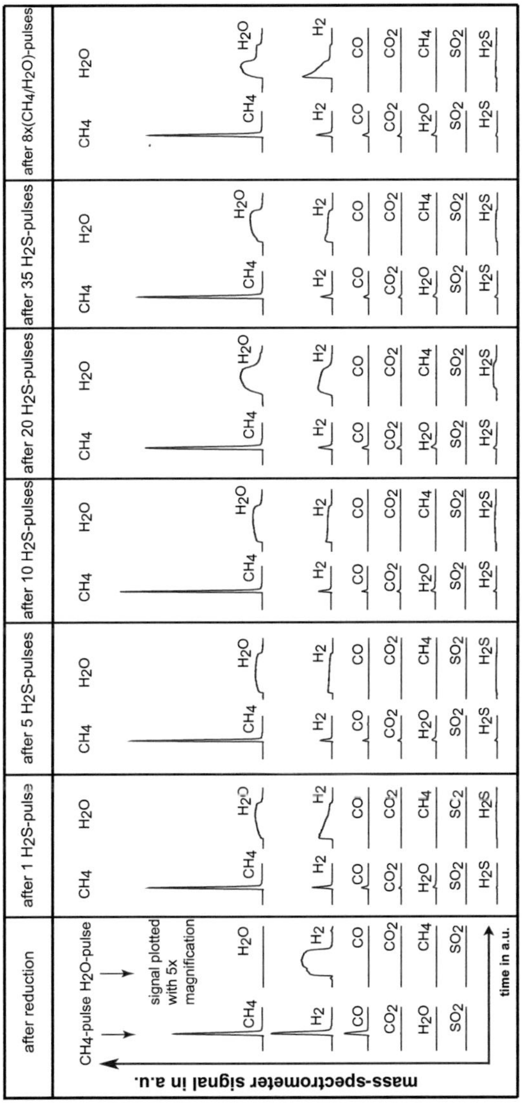

Figure 3.28: pulse experiments with the 1wt.-% Rh CeO$_2$ catalyst.

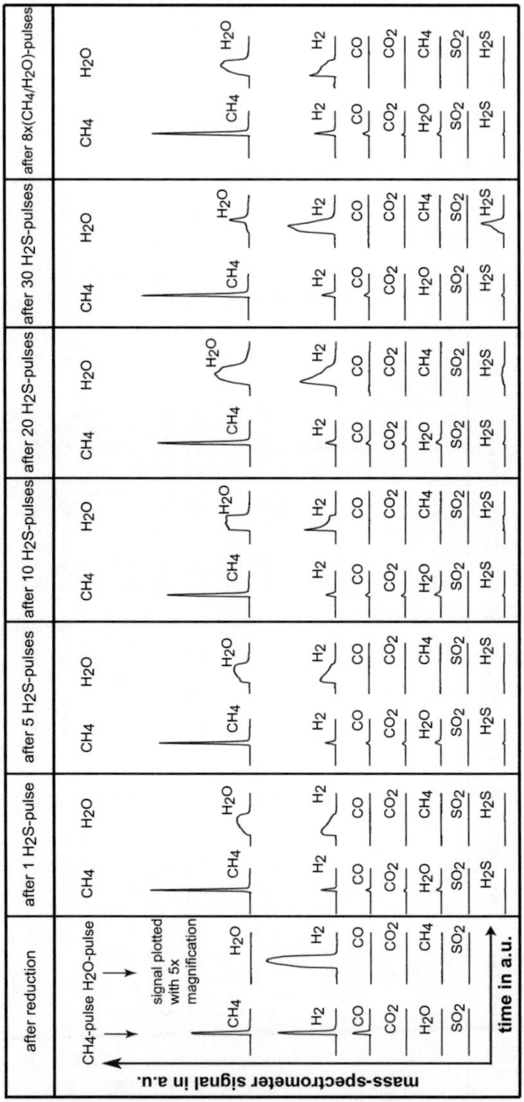

Figure 3.29: pulse experiments with the 1wt.-% Rh CGO (x_{Gd}=0.2) catalyst.

Figure 3.29 shows the result of the same transient pulse experiment as described above, only for the catalyst with a support containing 20% gadolinium (x_{Gd}=0.2). Considering the observation that the reforming reaction takes place mainly on the rhodium metal, no major differences are visible, the catalysts behave essentially alike - with one exception. Consistent with the observations above, the amount of H_2S during the methane water cycles (see after 35 pulses) is slightly higher, indicating that the sulfur is less strongly bound to the support, resulting in a more easily reversible sulfur adsorption. However after the sulfur free regeneration cycle, the activity of the catalyst was not significantly higher, but considering the minor difference measured earlier (see Figure 3.17), the difference in activity is probably not visible in the pulse experiments.

3.3.4. Summary

- the reducibility of the support results in the lattice oxygen participating in reactions with carbon, which results in an increased tolerance against coking

- sulfur and hydrogen sulfide interact with the support in a similar way as oxygen and water do

- for gadolinium doped ceria, the ceria - sulfur bond seems to be weaker than for pure ceria

- both the interaction of sulfur with rhodium and of sulfur with the support is reversible

- other sulfur compounds behave similarly, decomposing under reaction conditions to H_2, CO and sulfur adsorbed on the catalyst

3.4. Sulfur tolerance under reaction conditions

In Sections 3.3 and 3.2 several aspects of sulfur tolerance were investigated, allowing for the conclusion that the degree of deactivation likely depends on the sulfur adsorption equilibrium (Equation 3.5) as well as on the reaction of sulfur with the support (Equation 3.12). Both equilibria depend on the partial pressures of H_2S and of H_2 in the gas-phase. In the following, the influence of operational parameters on the activity of the Rh-CGO (x_{Gd} = 0.2) catalyst under typical steam-reforming conditions in a flow-though reactor is investigated in more detail and the results are interpreted using the insight stated above.

3.4.1. Sulfur adsorption equilibrium: Addition of hydrogen or nitrogen

Above, it was argued that the degree of activity depression depends on the sulfur adsorption equilibrium (Equation 3.5) and thus on the local hydrogen partial pressure in the reactor. This can be investigated directly by adding hydrogen to the reactor feed (see Figure 3.30 - all data points were obtained by measurements like the one shown in Figure 3.18, with the result at steady state being displayed in the graph). A large beneficial effect of H_2 addition on the activity is clearly visible.

For comparison, the same experiments were conducted with nitrogen as dilutant, and almost no effect of dilution can be detected (also shown in Figure 3.30). The improvement thus must be due to the influence of H_2 on the H_2S adsorption equilibrium. Also, on a side note, this result justifies the assumption that the reaction is first order in methane.

Given these results, it is clear that the sulfur adsorption equilibrium (Equation 3.5) is of crucial influence for the activity of the catalyst, with the partial pressure ratio of hydrogen sulfide to hydrogen governing the amount of accessible rhodium sites and thus the activity of the catalyst. Hydrogen addition therefore strongly enhances the attainable conversion when sulfur is present in the feed. As the amount of sulfur is fixed by the inlet conditions, the ratio is mainly governed by the hydrogen partial pressure, which in turn is determined by the conversion profile along the reactor length.

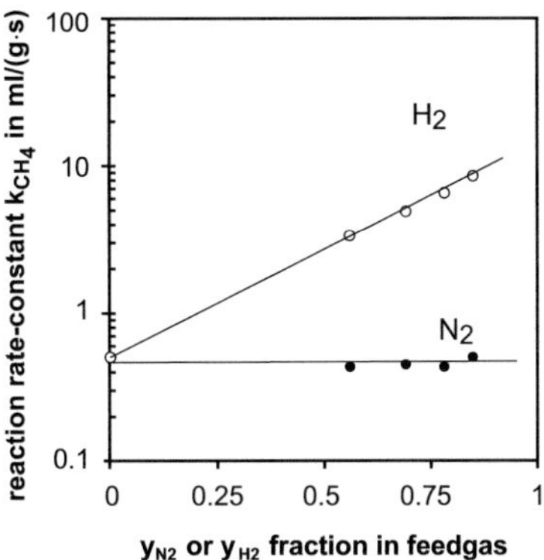

Figure 3.30: Comparison of the effects of dilution with N_2 and with H_2. T_R=800 °C, $p_{R,out}$=1.3 bar, τ_{mod}=0.13 (g·s)/ml, $\zeta = 3$, feedgas composition see Table 2.2, mixed with 14.3 mg/m³ COS and the indicated amount of N_2 or H_2

3.4.2. Influence of the amount of poison and of the feedgas composition

Figure 3.31 shows the typical feature of selective poisoning: a small amount of poison (H_2S in this case) is sufficient to lower the catalyst's activity significantly, and a further increase in poison loading in the feed lowers the catalyst's activity only slightly. A H_2S loading of about 35 mg/m³ is sufficient to suppress the methane conversion almost entirely. Considering the very low activity of the support (measured without sulfur in the feed), at this poison loading the methane conversion is fully deactivated.

Figure 3.31 also shows the depression of ethane and propane conversion. The higher reaction rate constant reflects the higher reactivity of ethane and propane compared to methane. It is interesting, however, that the conversion is not fully depressed for the high poison loading. As argued above, the catalytic activity of the rhodium metal for methane conversion is fully deactivated for $Y_S = 35$ mg/m³. From a reaction mechanism perspective, it is generally agreed that methane adsorbs as C-H_x (y=0–3) species on transition metals [18, 71, 76, 77, 78]. Higher hydrocarbons follow a similar pathway, sequentially forming C-H_x fragments and a remaining hydro-

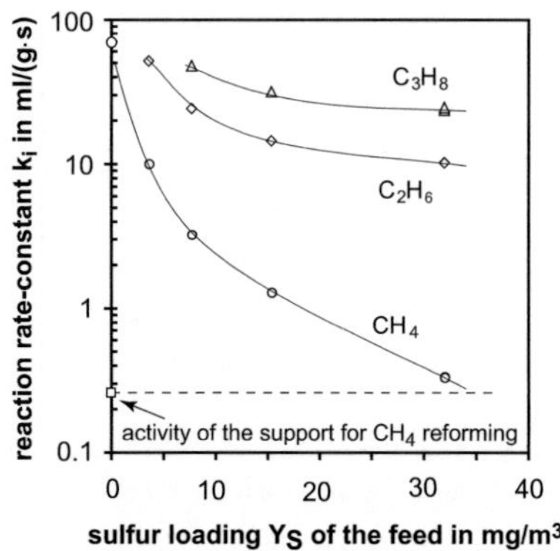

Figure 3.31: Reaction rate constant for methane, ethane and propane conversion with H_2S in the feedgas. T_R=800 °C, $p_{R,out}$=1.3 bar, τ_{mod}=0.13 (g·s)/ml, ζ = 3, feedgas composition see Table 2.2

carbon chain [8]. In both cases, the adsorbed C-H_x species reacts with an adsorbed OH group to form CO and H_2. If methane, ethane or propane all go through very similar surface intermediates, a surface poisoned for methane conversion should also be poisoned for ethane and propane conversion. Consequentially, the observed conversion of ethane and propane is likely due to the catalytic activity of the support or due to homogenous gas-phase reactions. Schädel made similar observations, concluding from experimental and modelling results that homogeneous gas phase reactions of higher hydrocarbons contribute to the overall conversion under typical steam reforming conversions [79].

As shown in Figure 3.32, the dehydrogenation products ethene and propene increasingly occur in the product gas for high loadings of sulfur in the feed, i.e. when the catalyst is inactive. Possible explanations for this behavior are that these products are formed by thermal gas-phase reactions or due to the catalytic activity of the support. When the rhodium surface is fully accessible, the dehydrogenated compounds likely react to CO, CO_2 and H_2 rapidly. When the catalyst is deactivated, this pathway is not possible, and the products can be measured in the reactor outlet. This

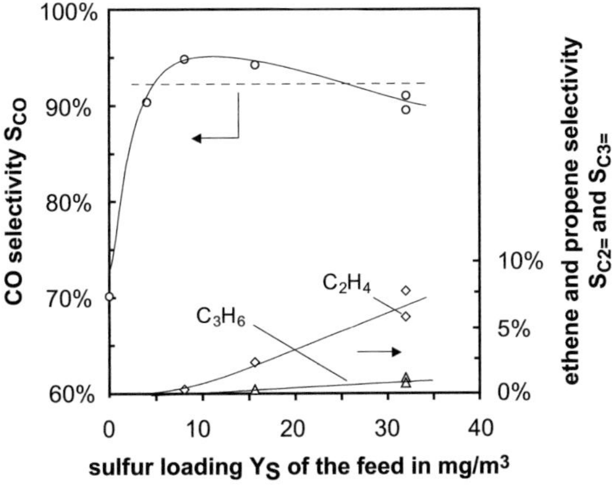

Figure 3.32: Selectivity towards CO, ethene and propene for the experiments shown in Figure 3.31.

explanation corresponds to trends observed for the autothermal reforming of liquid hydrocarbons, where too short contact times lead to the detection of olefines in the product [80, 81].

In any case, only a small fraction of ethane and propane reacts to the dehydrogenation products: the product selectivity from ethane to ethene and from propane to propene is 16% and 3% respectively (Note: the selectivity given in Figure 3.32 is related to the total amount of reacting carbon). In consequence, the reforming of ethane and propane and likely any other higher hydrocarbons takes place to some extent regardless of the sulfur added to the feed. This plays an important role regarding the activity of the catalyst, as hydrogen in the gas phase lowers the degree of deactivation (see 3.4.1). Indeed, when natural gas as feedgas is replaced by pure methane under otherwise equal conditions, a lower methane conversion is measured, see Figure 3.33. Interestingly, under sulfur-free conditions, the methane conversion with methane only fed is higher compared to the natural gas case, probably due to the competition by the higher hydrocarbons for rhodium surface sites (see Figure 3.17). This result therefore indicates that the positive effect on the activity due to the additional hydrogen in the

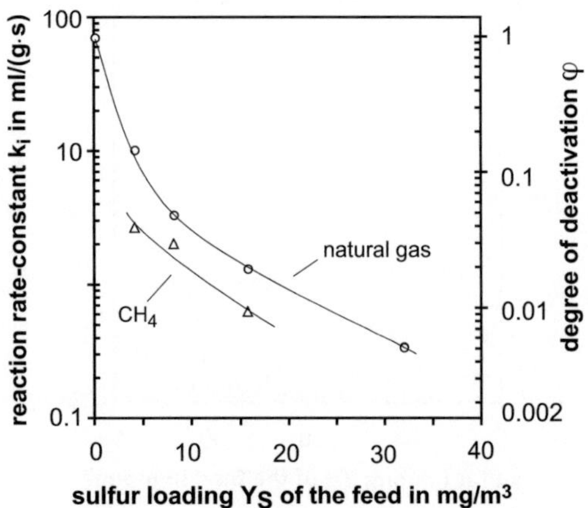

Figure 3.33: Reaction rate constant for natural gas and methane as feedgas. T_R=800 °C, $p_{R,out}$=1.3 bar, τ_{mod}=0.13 (g·s)/ml, ζ = 3, feedgas composition pure methane or natural gas, see Table 2.2.

gasphase (due to the reacting higher hydrocarbons) outweighs the negative one by the competitive adsorption of higher hydrocarbons.

Figure 3.32 also shows that the CO selectivity rises dramatically when sulfur is added to the feed. Whether the CO selectivity really undergoes a maximum or remains constant (both shown with grey lines in Figure 3.32) cannot be determined due to the rather low conversion and the thus rather high experimental error in determining the CO selectivity. Irrespective of the exact nature, the rise in CO selectivity corresponds to the observation that the shift reaction is strongly affected by the addition of sulfur to the feed, as discussed in relation to Figure 3.18.

3.4.3. Influence of the type of sulfur compound

In natural gas, other sulfurous molecules in addition to H_2S are present. For example, carbonoxysulfide (COS) is a common sulfurous compound encountered in natural gas. Further, as mentioned above in relation to

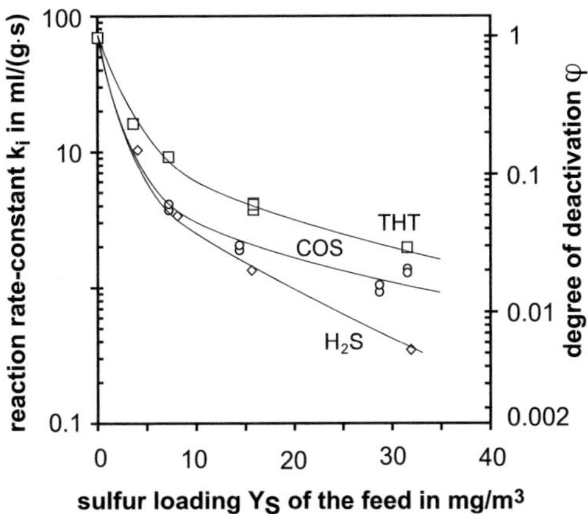

Figure 3.34: Comparison of the deactivation upon addition of H$_2$S, COS and THT. T_R=800 °C, $p_{R,out}$=1.3 bar, τ_{mod}=0.13 (g·s)/ml, ζ = 3, feedgas composition see Table 2.2 with sulfur component added as indicated.

Figure 3.19, tetrahydrothiophene (THT) a typical natural gas odorant used in Europe, is to be expected.

As shown in Figure 3.34, COS and especially THT have a weaker effect on the steam-reforming activity compared to H$_2$S.

The type of sulfur compound thus plays a significant role in sulfur poisoning. This is somewhat unexpected, as under steam reforming conditions, all mercaptans and sulfides present in the distributed natural gas are unstable and should decompose to yield adsorbed sulfur (and H$_2$, CO and CO$_2$), see Figure 3.27. However, since adsorbed sulfur is the acting poison, both THT and COS first need to be decomposed to affect the reforming reaction rate, while H$_2$S readily adsorbs disassociatively. Likely, COS and THT are decomposed more slowly than H$_2$S. This probably leads to more favorable hydrogen concentration profiles, especially at the entrance of the reactor, resulting in lower degrees of deactivation.

During most of the experiments, only H$_2$S was observed in the reactor effluent, regardless of the type of sulfur compound added to the feed. In

Table 3.6: Sulfur balance error for several experiments with different sulfur components and loadings

sulfur loading in the feed in mg/m^3	H_2S	COS	THT
	sulfur balance error in %		
8	16.1	8.5	-34.7
17	-1.9	-8.7	-5.0
33	4.2	0.0	-12.7

some cases however, COS was also found as product, which is likely due to the equilibrium below:

$$CO_2 + H_2S \quad \rightleftharpoons \quad COS + H_2O \tag{3.13}$$

The sulfur balance error was mostly low (see Table 3.6), thus both THT and COS are likely converted (within equilibrium limitations) to H_2S under reaction conditions.

3.4.4. Influence of the operating conditions

While discussing the results presented in this section, it was argued that the sulfur adsorption equilibrium (Equation 3.5), and thus the partial pressures of H_2S and H_2 in the gas-phase determine the degree of activity depression. Following that argument, it should be possible to adjust the operating conditions to allow for higher activity. In the following, experiments revealing this influence of the operating conditions are presented.

Steam-to-carbon ratio ζ It would be expected that higher steam-to-carbon ratios lead to a lower degree of deactivation (i.e. higher activity) under otherwise equal conditions, since the higher steam-to-carbon ratio favors the formation of hydrogen due to the water-gas shift equilibrium. This is indeed observed, as shown in Figure 3.35. Judging from the results, a steam-to-carbon ratio of about $\zeta=3$–4 appears to be optimal, as more water vapor addition is energy intensive due to the evaporation needed, but does not increase the activity significantly. In general, however, the optimal steam-to-carbon ratio most likely has to be determined individually for each fuel-cell system design.

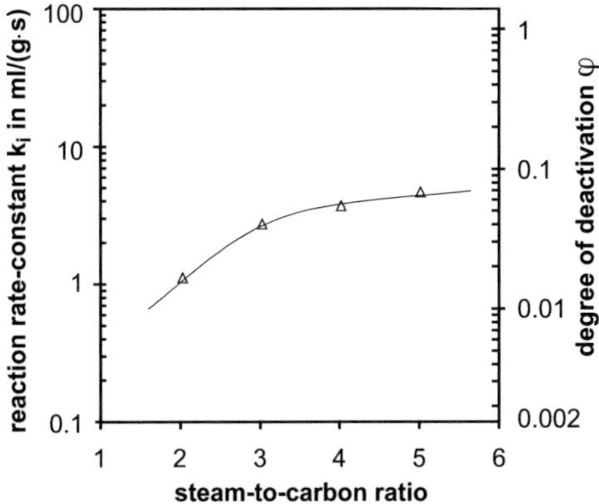

Figure 3.35: Observed activity depression for different steam-to-carbon ratios. 7.6 mg/m³ COS addition, T_R=800 °C, $p_{R,out}$=1.3 bar, τ_{mod}=0.13 (g·s)/ml, feedgas composition see Table 2.2.

It has to be noted that at low steam-to-carbon ratios, the danger of carbon deposition ("coking") is aggravated by the addition of sulfur (i.e. the activity depression) especially at the entrance of the bed. In the experiments, an increase in pressure drop over the catalyst bed was noticeable for the runs at $\zeta = 2$, which was probably due to carbon deposition. As discussed in Section 2.3.2, the data points were derived assuming no carbon balance error. However, as the nitrogen content was measured routinely, the carbon balance error can be calculated (via the internal standard method) and amounted to about 7 %. This, together with the slow increase in pressure drop indicates that there is slow carbon deposition taking place. The pressure drop returned to the values measured at the beginning of the experiment after runs at higher steam-to-carbon ratio or after treatment with a N_2/H_2O flow (also see 3.20), which further confirms that the observed effects are likely due to carbon deposition. These results also show that the carbon balance alone is not a good indicator of carbon deposition, as an error of 7 % is relatively small and close to experimental errors obtained regularly.

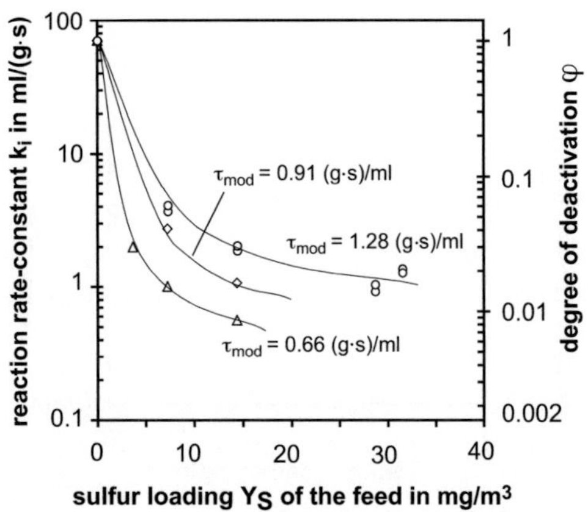

Figure 3.36: Observed activity depression for different rates of addition and residence times for COS as sulfur compound. T_R=800 °C, $p_{R,out}$=1.3 bar, $\zeta = 3$, feedgas composition see Table 2.2.

Residence time τ Because longer residence times result in higher steam-reforming conversion at the outlet of the reactor, the hydrogen partial pressures reached in the reactor are generally higher. Consequentially, the adsorption equilibrium is shifted towards empty (and therefore active) rhodium surface sites, which explains the dependence of the observed activity on the residence time (see Figure 3.36).

Temperature T_R If the sulfur adsorption equilibrium plays a decisive role for the activity of the catalyst, the influence of the reaction temperature must be twofold. In addition of the well known exponential dependence of the activity with temperature (Arrhenius' expression), there is a temperature dependence of the sulfur adsorption equilibrium. It is therefore expected that there is a very strong dependence of the activity on the temperature, which is confirmed experimentally (see Figure 3.37). The activation energy measured for the sulfur free case is 95 kJ/mol, which compares well with the values measured for pure methane (83–93 kJ/mol, see Table 3.5(a)).

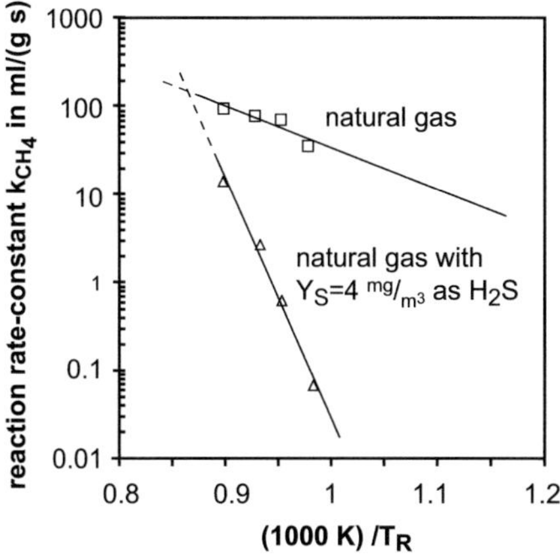

Figure 3.37: Influence of the reactor temperature T_R on the observed activity with and without H_2S addition (Y_S=4 mg/m³). $p_{R,out}$=1.3 bar, $\zeta = 3$, feedgas composition see Table 2.2.

As obvious from the graph, the *effective* activation energy for the case with Y_S=4 mg/m³ in the feedgas is determined to be much larger. As mentioned above, the temperature dependence in this case is a combination of influences, and thus the actual number (522 kJ/mol) is only valid for this particular set of operating conditions and thus is of no general value. Nevertheless, the result shows that the temperature, in addition to the well known impact on the reaction rate, has a very strong influence on the catalyst's degree of deactivation. Moreover, extrapolation of the trend allows to estimate that at 890 °C, the addition of sulfur should not cause any harmful effect to the catalyst's activity in this case. Unfortunately, a limitation of the experimental setup (for the oven temperature) did not allow for a validation of this matter.

3.4.5. Summary

- The observed tolerance towards sulfur depends on the operating parameters such as the steam-to-carbon ratio, the residence time and the feedgas composition.

- The effect of sulfur is non-linear, corresponding to selective poisoning.

- Generally, all factors leading to a high hydrogen partial pressure are beneficial for the catalyst's activity under sulfur containing feed. Out of those factors, the reactor temperature has the largest impact and both the steam-to-carbon ratio and the residence time have a somewhat smaller influence.

- Among the factors outside of operational control, the type of sulfur component has a large influence, THT is much less severe of a poison compared to H_2S and COS. The feedgas composition in regards to the hydrocarbons is not quite as important, although the presence of higher hydrocarbons is beneficial.

- In addition to the tolerance towards sulfur components in the feed, the sulfur components are fully converted. H_2S is the main sulfurous reaction product.

3.5. Modeling the reactor without and with H$_2$S in the feed

Supplementing the experiments shown in the previous sections, a series of experiments using the natural gas composition shown in Table 2.2 was carried out (see Appendix **??**). The parameter range covered three steam-to-carbon ratios (ζ=1.5, 2.0, 3.0) and seven feed gas volume flows ($^V\phi$=5, 7, 9, 11, 13, 15, 20, 25 l/h (NTP). In the following, the modeling of these experiments with the 'full model' described earlier is discussed. Thereafter, the model is modified to account for the reversible effect of sulfur (in the form of H$_2$S) in the feed. Using the determined parameters and the model, the results are extrapolated to conditions appropriate for a fuel cell system.

3.5.1. Sulfur free operation

In addition to the , two assumptions were made to simplify the modeling.

Propane, the butanes and pentane were assumed to be fully converted at the inlet of the reactor. The CO and H$_2$ concentration at the inlet of the reactor therefore are not zero, but correspond to the full conversion of the amounts of higher hydrocarbons in the feed. Equally, the amount of water in the inlet is reduced by the water already consumed by the reaction of the higher hydrocarbons. The shift reaction, in contrast, was assumed to begin in the catalyst bed only, so the concentration of CO$_2$ is always zero at the inlet. As the higher hydrocarbons are reacting much faster than methane, as only a small amount of higher hydrocarbons are present in the reaction feed and as some homogeneous reaction of the higher hydrocarbons is likely to take place during the heating of the feed prior to the entrance in the catalyst bed, the above simplification has only little impact on the calculated concentration profiles, and ultimately on the fitted reaction rate constants for methane and ethane.

Secondly, it was observed during the use of the model that the pressure profile along the reactor is very close to linear. Simplifying the model to yield a linear pressure profile between the (known) inlet and outlet pressures resulted in a significant gain in calculation speed, thus this approach was implemented. The validity and impact of this simplification was tested by calculating the pressure profiles using both methods. The largest difference (between the quadratic profile calculated by the Ergun equation and the linear approximation) was 0.011 bar, with an inlet pressure of 1.68 bar, and an outlet pressure on 1.3 bar, which is sufficiently small to be neglected.

In Section 2.3.2 the equations forming the basis of the full model were described. For implementation of the model, rate expressions are required. Three reactions in parallel are considered, methane and ethane steam reforming (Reaction 3.3, denoted here as "1" and analogous for ethane, denoted as "2") and water-gas shift equilibrium (Reaction 3.4, denoted here as "3").

For the steam reforming reactions a power-law rate expression is used for simplicity at first:

$$r_1 = k_1 (c_{CH_4})^{\alpha_1} (c_{H_2O})^{\beta_1} \tag{3.14}$$

$$r_2 = k_2 (c_{C_2H_6})^{\alpha_2} (c_{H_2O})^{\beta_2} \tag{3.15}$$

Under reaction conditions, the conversion in the water-gas shift reaction might be restricted by equilibrium, therefore, a modified power-law rate expression is used:

$$r_3 = k_3 (c_{CO})^{\alpha_3} (c_{H_2O})^{\beta_3} \left(1 - \frac{K_p}{K_{p,eq}} \right) \tag{3.16}$$

in which

$$K_p = \frac{(p_{H_2})(p_{CO_2})}{(p_{CO})(p_{H_2O})} \tag{3.17}$$

and $K_{p,eq} = K_p$ at equilibrium. $K_{p,eq}$ can be determined from thermodynamic data (in this case extraced from the ASPEN Engineering suite software package) in the temperature range of interest. The last factor on the right hand side of Equation 3.16 is approaching unity for gas compositions far from equilibrium, while for compositions close to equilibrium the factor and thus the reaction rate approaches zero.

With the above model, a total of nine parameters are needed to describe the system (note that the temperature is not varied, but kept at 800 °C). To estimate the parameters giving the smallest error, optimizer program ('leasqr') was allowed to vary all 9 parameters simultaneously. Figure 3.38 shows parity plots for the three to be predicted variables methane conversion, ethane conversion and CO_2 selectivity. As can be seen, the model predicts the measured values quite well. Figure 3.38 also shows the standardized residuals, calculated as

$$SR = \frac{X_{CH_4,measured,n} - X_{CH_4,predicted,n}}{X_{CH_4,measured,n}} \tag{3.18}$$

in the case of the methane conversion. The SR values for the ethane conversion and the selectivity were calculated analogously. With the exception of a slightly higher error at small methane conversions, no systematic correlation of the residual is observed.

Table 3.7 shows the fitted parameters along with their standard deviation σ and the residual sum of squares, RSS, calculated in the case of the methane conversion as

$$RSS = \sum_n \left(X_{CH_4,measured,n} - X_{CH_4,predicted,n} \right)^2 \tag{3.19}$$

for all n measurements. The RSS values for the ethane conversion and the selectivity were calculated analogously.

The parameters obtained are generally in the range of published values, see [82] for a review of reforming kinetics. The standard deviations obtained however are high, sometimes higher than the fitted parameter values themselves. This indicates that there are strong cross-correlations of the parameters, and that the model likely contains too many (or unadequate) parameters.

To improve the model and to reduce the number of required model parameters, various other formulations for the rate equations can be tried. First, the resulting reaction order of CO in the water gas shift reaction, α_3 was determined to be very close to zero, thus the rate equation was simplified such that the reaction order for CO was set to zero. The fitted parameters for this set of rate equations are also shown in Table 3.7. As expected, the parameter values do not differ significantly, the residuals are slightly higher, but the standard deviations of the parameters are slightly smaller. Further reducing the number of parameters, two more reactions orders were "rounded" and fixed. This results in again slightly higher residuals, but the standard deviations now are all smaller than the values of the parameters, yielding a valid model.

Fundamental studies of Wei and Iglesia [71] resulted in the conclusion that methane reforming should be first order in methane and zeroth order in water. In Table 3.7, the reaction order of water in the ethane and methane reforming are small, so reaction rate equations being first order in the hydrocarbon and zeroth order in water were tried, see Table 3.7. While the inaccuracy in the predicted values increases due to the above simplifications, the resulting errors are still relatively small. This gives justification for the simplified treatment described in section 2.3.2.

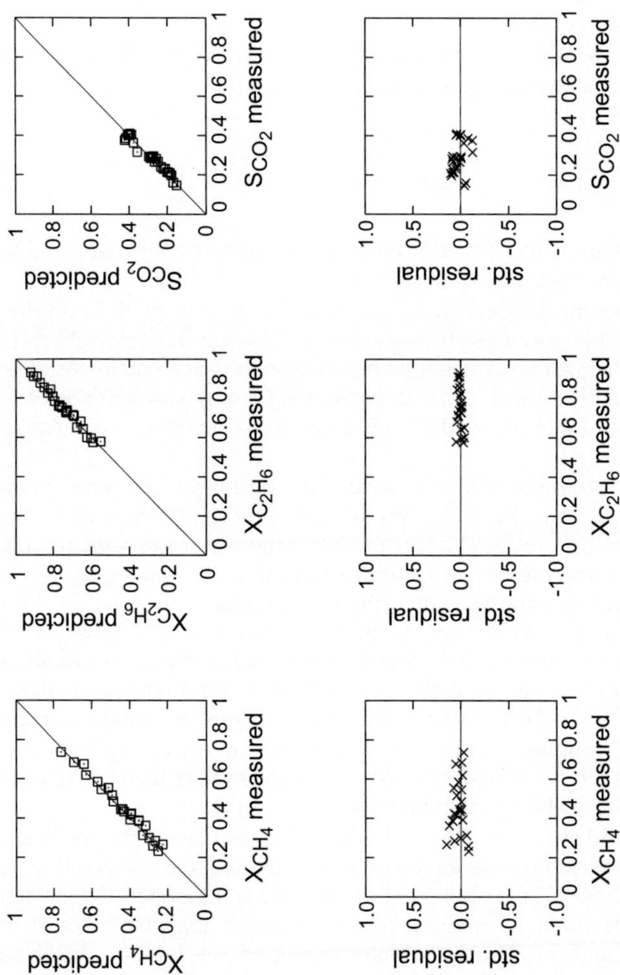

Figure 3.38: parity plots for the model with variable power-law kinetics. Parameters of case 3 in Table 3.7 were used.

Table 3.7: Fitting results for the experiments using power-law reaction rate expressions 3.14, 3.15 and 3.16. Note: units for the rate constants k_i depend on the exponents: $[k_i] = \dfrac{ml^{(\sum_i \alpha_i + \sum_i \beta_i)}}{gs(mmol)^{(\sum_i \alpha_i + \sum_i \beta_i - 1)}}$

	methane reforming			ethane reforming			shift reaction		
	k_1	α_1	β_1	k_2	α_2	β_2	k_3	α_3	β_3
case 1: simultaneous fitting of all 9 parameters									
value	57.27	0.70	0.12	51.12	0.92	0.35	3.88	0.02	1.39
σ	49.51	0.59	0.56	217.4	0.97	1.33	2.69	0.25	0.30
R^2	0.9870			0.9597			0.9868		
RSS	61.2			49.9			217.7		
case 2: simultaneous fitting of 8 parameters, fixing β_3									
value	58.80	0.73	0.09	51.44	0.92	0.35	4.00	0	1.39
σ	81.78	0.39	0.39	8.95	0.02	0.07	1.31	–	0.13
R^2	0.9852			0.9870			0.9577		
RSS	70.8			43.4			249.2		
case 3: simultaneous fitting of 6 parameters, fixing β_1, α_2 and α_3									
value	70.13	0.75	0	61.32	1	0.37	4.21	0	1.36
σ	17.99	0.38	–	5.58	–	0.05	1.53	–	0.15
R^2	0.978			0.9809			0.9554		
RSS	113.2			91.3			368.9		
case 4: 'forced' first and zeroth reaction orders for methane and ethane									
value	56.85	1	0	126.47	1	0	5.03	0	1.29
σ	0.88	–	–	1.97	–	–	0.41	–	0.04
R^2	0.9476			0.9513			0.9234		
RSS	347			327			640		

While the above presented power-law based equation set is probably sufficient for the extrapolation to realistic conditions for fuel cell systems, it is still worthwhile to explore wether better parametrisations can be found for the methane and the ethane reforming. While there are various theoretical issues questioning the validity of Langmuir-Hinshelwood (LHW) type rate equations, they still have been proven to be of great practical value (see e.g.[83] for a discussion). Below, a set of LHW rate equations for the hydrocarbon conversion is derived. Regarding the water gas shift reaction on ceria supported noble metals, Gorte and Zhao reviewed the current knowledge, concluding that no undisputed reaction mechanism exists and that multitude of different reaction rate equations can be found in literature [75]. As the parametrisation given in Equation 3.16 give satisfactory results, no further attempt at finding a more suitable rate equation was undertaken.

To describe the methane conversion, it is assumed that the reaction rate is proportional to the methane surface coverage of the rhodium, but independent of the amount of water present. This is consistent with Wei and Iglesia's studies , mentioned above, because this yields first order in methane at the high dilution conditions employed by Wei and Iglesia [71].

$$r_1 = k_1 \theta_{Rh,CH_4} \tag{3.20}$$

the surface coverage θ_{Rh,CH_4} can be represented by a Langmuir adsorption isotherm, with

$$\theta_{Rh,CH_4} = \frac{K_{ads,CH_4} c_{CH_4}}{1 + K_{ads,CH_4} c_{CH_4}} \tag{3.21}$$

Combining Equations 3.20 and 3.21 yields

$$r_1 = \frac{k_1 K_{ads,CH_4} c_{CH_4}}{1 + K_{ads,CH_4} c_{CH_4}} \tag{3.22}$$

Using Equation 3.22, the methane conversion could be modeled sufficiently accurate, however the ethane conversion, modeled analogously, could not be represented adequately. Beretta et al. investigated a Rh/Al_2O_3 catalyst and proposed to additionally consider inhibition by CO adsorption [84]. Adding this inhibition to Equation 3.22 yields

$$r_1 = \frac{k_1 K_{ads,CH_4} c_{CH_4}}{1 + K_{ads,CH_4} c_{CH_4} + K_{ads,CO} c_{CO}} \tag{3.23}$$

for methane and

$$r_2 = \frac{k_2 K_{ads,C_2H_6} c_{C_2H_6}}{1 + K_{ads,C_2H_6} c_{C_2H_6} + K_{ads,CO} c_{CO}} \tag{3.24}$$

for ethane.

Earlier, the results of the power-law fitting (see Table 3.7) showed that the reaction order of ethane is close to one. This together with the fact that only little ethane is present in the feed natural gas, indicates that likely $K_{ads,C_2H_6} c_{C_2H_6} \ll K_{ads,CO} c_{CO}$, and that therefore Equation 3.24 can be further simplified to

$$r_2 = \frac{k_2 K_{ads,C_2H_6} c_{C_2H_6}}{1 + K_{ads,CO} c_{CO}} = \frac{k_2^* c_{C_2H_6}}{1 + K_{ads,CO} c_{CO}} \tag{3.25}$$

Using the CO adsorption constant determined by Beretta et al ($K_{ads,CO}=$ 211 1/atm, and $K_{ads,CO} \approx 0.737\,\mathrm{m^3/mol}$ extrapolated to the conditions here) as starting value, the data here can be fitted well, see Figure 3.39. Even more, the fitted value for $K_{ads,CO}$ is only marginally different from the published value, see Table 3.8.

The advantage of the LHW type set of reaction rate expressions is clearly that at least one of the parameters is externally confirmed - nevertheless, this model also uses 6 fitted parameters. Analyzing the standard deviations for the parameters and the quality of fit (represented by R^2 and RSS), the power-law set of equations (Table 3.7, case 3) seem to fit the data slightly worse while giving tighter confidence intervals. In conclusion both parameter sets - Case 3 in Table 3.7 and Case 2 in Table 3.8 give valid models of the experimental data, and more measurements would be required to discriminate between both models.

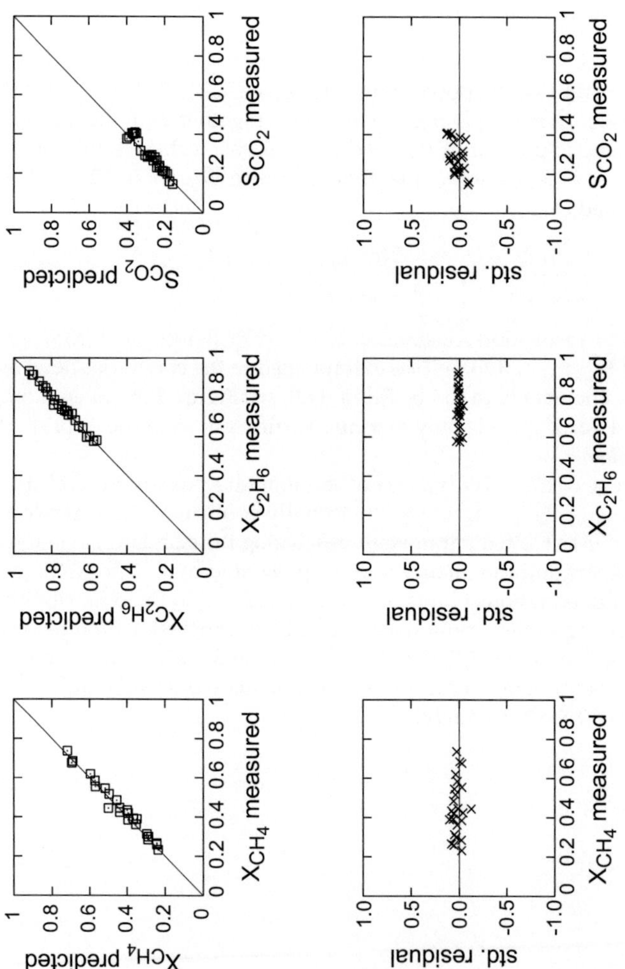

Figure 3.39: parity plots for the model with Langmuir-Hinshelwood type rate equations 3.23 and 3.25 for methane and ethane. The power-law expression (equation 3.16 with $\alpha_3=0$) was used for the water gas shift reaction. See Table 3.8

Table 3.8: Fitting results for the experiments using LHW reaction rate expressions 3.23, 3.25 and using the power-law expression 3.16 for the shift reaction.

	k_1 $ml/(gs)$	k_2^* $ml/(gs)$	K_{ads,CH_4} m^3/mol	$K_{ads,CO}$ m^3/mol	k_3 $\dfrac{ml^{\beta_3}}{gs(mmol)^{(\beta_3-1)}}$	β_3 $(\alpha_3 = 0)$
			case 1: simultaneous fitting of all 6 parameters			
value	330.79	246.61	0.613	0.789	3.802	1.398
σ	251.06	33.79	0.899	0.328	1.396	0.1556
R^2	CH$_4$	0.9767	C$_2$H$_6$	0.9881	Shift	0.9265
RSS		167.6		40.59		313.71
			case 2: simultaneous fitting of 5 parameters, keeping $K_{ads,CO}$ fixed			
value	336.78	247.16	0.602	0.789	3.789	1.399
σ	138.58	4.96	0.391	–	1.267	0.14955
R^2	CH$_4$	0.9769	C$_2$H$_6$	0.9881	Shift	0.9282
RSS		119.44		40.12		307.22

3.5.2. Operation with H_2S in the feed

It was shown in section 3.4 that the effective rate constant k_{CH_4} is positively influenced by high hydrogen partial pressures and, as expected, negatively by sulfur in the feed. Mechanistically, this was explained by the sulfur adsorption equilibrium on rhodium (Reaction 3.5).

One approach to model the effect of sulfur poisoning is to relate the effective rate constant to the intrinsic rate constant under sulfur free operation multiplied by an activity factor φ, which depends on the rhodium surface coverage with sulfur $\theta_{Rh,S}$ [85].

$$k_1 = \varphi \cdot k_{1,noS} \; ; \varphi = f(\theta_{Rh,S}) \tag{3.26}$$

The rhodium surface coverage can be described by a Langmuir type adsorption isotherm:

$$\theta_{Rh,S} = \frac{K_{ads,H_2S} \frac{p_{H_2S}}{p_{H_2}}}{1 + \left(K_{ads,H_2S} \frac{p_{H_2S}}{p_{H_2}} \right)} \tag{3.27}$$

using the rhodium surface coverage $\theta_{Rh,S}$, or rather the fraction of empty surface sites, $(1-\theta_{Rh,S})$, the activity factor can be calculated (as proposed by Rostrup-Nielsen [85]) by

$$\varphi = (1 - b\theta_{Rh,S})^c \tag{3.28}$$

in which b is in fact the product of two factors, one indicating the maximum achievable surface coverage and the other indicating how many surface sites are deactivated by one sulfur atom. The other parameter, c, is related to the number of free surface sites needed for the steam reforming reaction to proceed.

To fit the parameters b and c, the measurements with several H_2S loadings and three residence times were used, using both the power-law and the LHW model as basis. Because it was shown that there is likely some homogeneous gas-phase reaction of the higher hydrocarbons (see discussion of Figure 3.31), the ethane conversion was not used to fit the parameters b and c. Equally, under sulfur loaded feed the shift reaction is strongly inhibited (see section 3.2.2) so the shift reaction rate was set to zero for the cases with sulfur in the feed.

First, it can be noticed by looking at the fitting results shown in Table 3.9 (also see Figure 3.40 for parity plots), that the parameter b is close to 1 for

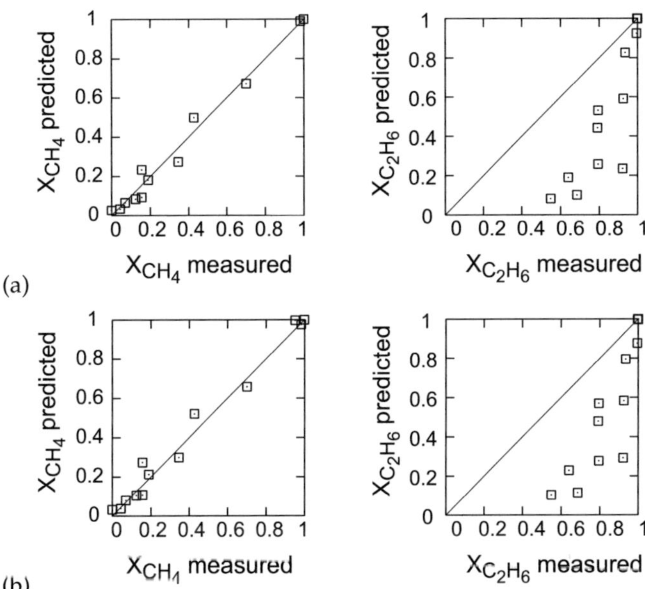

Figure 3.40: parity plots for the model with sulfur in the feed, with b and c fixed to 1. (a) for power-law model as basis, (b) LHW model as basis. See Table 3.9, cases 1b and 2b.

Table 3.9: Fitting results for the experiments with sulfur

	K_{ads,H_2S}	b	c
	–	–	–
	case 1: power-law model (case 3 in Table 3.7) as basis		
value	$1.842\ 10^6$	0.984	1.098
σ	$17.00\ 10^6$	0.123	2.538
R^2	CH$_4$	0.9839	
RSS		439	
	case 2: LHW model (case 2 in Table 3.8) as basis		
value	$1.956\ 10^6$	0.999	0.999
σ	$12.56\ 10^6$	1.057	1.814
R^2	CH$_4$	0.9820	
RSS		333	

both the power-law and the LHW model as basis. For nickel, it was shown that the maximum surface coverage is approximately 0.5 ML (monolayer) and that one sulfur atom deactivates 2 nickel atoms, thus for nickel, $b=1$ [86]. For rhodium, however, it was shown that a surface coverage of at least 0.75ML can be achieved [87]. Consequentially, one sulfur atom can only poison 1.33 or less rhodium surface atoms. This result is in agreement with density functional theory studies by Zhang et al. [88], who showed that the effect of adsorbed sulfur on Rh <111> surfaces is very local, with only a small influence seen for nearest-neighbour atoms.

Further, the fitting parameters show that the parameter c is close to 1 for both models as well, indicating that only one rhodium atom is required for the methane steam reforming reaction. This is distinctly different from the nickel case, on which a value of 3 is found (indicating an 'ensemble' of 3 nickel surface atoms is required for the reaction) [86]. With the ceria containing carrier, lattice oxygen participates in the reaction. This makes simultaneous adsorption of water on the rhodium surface is not required, which might explain why only a smaller ensemble is required for the reaction to proceed (see Figure 3.23 for a schematic).

From a reaction mechanism perspective, the value of exactly one site only is rather unlikely, as this would indicate associative rather than dissociative adsorption. However, the simple model using a number of otherwise equal adsorption sites is likely too simplistic, as different crystallographic planes of rhodium likely exhibit different activity. The value of the parameter c should therefore not be interpreted quite so direct as number of adsorption sites but rather as an indication of the required rhodium surface area.

The above result indicates that for the catalyst under investigation significantly less metal surface area is required compared to a typical nickel-alumina steam reforming catalyst. This is plausible, as the support participates (via lattice oxygen) in the steam reforming reaction. This means that water does not have to adsorb on the rhodium, and therefore less rhodium surface is required for the reforming reaction to proceed, and thus making the impact of sulfur adsorption on rhodium less severe. This could be one of the reasons for the observed sulfur tolerance.

The numerical value of the equilibrium constant K_{ads,H_2S} is close to the value given by Rostrup Nielsen for nickel ($1.2\ 10^6$). The fact that the fitted value is slightly higher than the value for nickel is surprising, because it would be expected that the adsorption is slightly less strong on the nobel metal. However, McCarty and Wise [89] showed that the thermodynamics sulfur chemisorption on nickel is dependent on the surface coverage, which makes the results obtained by the Langmuir adsorption isotherm inaccurate (because the surface coverage not only depends on the gas phase partial pressure but also on the surface coverage itself). The equilibrium constant should therefore probably also be seen more as a fitting parameter, without placing too much emphasis on the obtained numerical value.

Figure 3.40 not only shows the parity plots for methane, but also for ethane. As already observed in relation to Figure 3.31, ethane is likely converted in a gas-phase reaction as well, which leads to large underprediction of the ethane conversion. As mentioned earlier, the water gas shift reaction is almost entirely deactivated, so no attempt to model the conversion of the water gas shift reaction was undertaken.

Figure 3.41 shows a comparison between the measured values and the trend calculated by the model in a format similar to the plots presented earlier. Generally, the fit between the model and the experimental values is quite good, especially considering that the rate constants are plotted on a logarithmic axis. Nevertheless, as a conclusion from the above discussion, and further noting the uncertainty associated with the parameter estimation, it becomes clear that while Equation 3.28 is yielding a quite well fitting model of the sulfur poisoning, the interpretation of the parameters b, c, and K_{ads,H_2S} in a physical sense is more difficult.

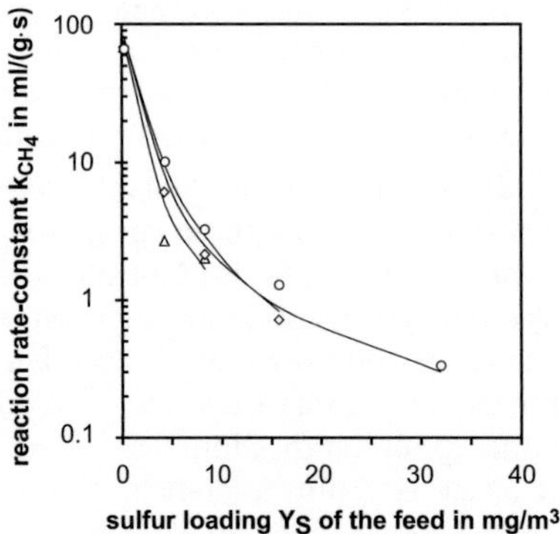

Figure 3.41: Observed activity depression for different rates of addition and residence times for H_2S as sulfur compound. Same conditions as in Figure 3.36. Lines are generated with the LHW model, see Table 3.9, case 2b.

Nevertheless, the interpretation that H_2S adsorption on rhodium causes a much more local effect compared to nickel, and that less noble metal surface area is required for the steam reforming reaction to proceed is intriguing. Compared to nickel catalysts, these effects would lead to much higher activity factors φ at the high sulfur surface coverages $\theta_{Rh,S}$ observed here, which may explain the observed sulfur tolerance of the catalysts under investigation.

Chapter 4.

Economic evaluation

Concluding from the experimental work presented in section 3.4, the rhodium catalyst supported on gadolinium doped ceria with $x_{Gd} = 0.2$ provides a useful and sustained sulfur tolerance. Several size classes of fuel cell systems for domestic use are envisioned, ranging from $P_{el.} = 1\,kW$ to $5\,kW$ electrical output. Further, combined heat and power stations for larger areas typically provide $250\,kW$ electrical output. The comparison is carried out for these three fuel cell system sizes.

In the following, a simple evaluation is presented to explore whether the sulfur tolerant process option is economically viable compared to a process scheme with adsorbent for natural gas desulfurization and a nickel catalyst for steam reforming. In such a process, the adsorbent bed has to be replaced and serviced periodically, which causes yearly expenses. On the other hand, there is a larger upfront investment for the sulfur tolerant processing scheme, because the rhodium catalyst is significantly more costly than the nickel catalyst.

Typically, one would have to conduct a process design exercise for both options and use the resulting equipment design data to estimate the required capital expenses for both options. The process line-up and the equipment design itself however, are not very different for both options under consideration. Table H.1 in Appendix H shows the prices of relevant raw materials for the employed catalysts. It is obvious that rhodium is by far the most valuable component used and therefore, it is assumed for simplicity that the additional cost for the sulfur tolerant process option is equal to the cost of the rhodium contained in the catalyst.

To estimate the amount of catalyst needed for the three system sizes under consideration, the following assumptions are made:

- The fuel cell stack (PEM type has been assumed) has an electrical generation efficiency of $\eta_{el.} = 45\,\%$ (based on electrical output and LHV of the hydrogen feed).

- The same natural gas as used in the experiments is assumed (see Table 2.2).

- In the reformer, a methane conversion of 98% is required. The same reaction conditions as in the experiments are used, i.e. reactor temperature 800 °C, a steam to carbon ratio $\zeta = 3$, and a feed pressure p=2 bar.

Given the desired power generation, the amount of hydrogen gas to be fed can be calculated:

$$^N\Phi_{H_2} = \frac{P_{el.}}{\eta_{el.} H_{l,H_2}} \tag{4.1}$$

in which H_{l,H_2} is the lower heating value of hydrogen (240 MJ/mol).

Using the model and the parameters as described in section 3.5, the amount of catalyst mass required for such a hydrogen flow can be calculated. Table 4.2 shows the required catalyst mass for the three system sizes under consideration.

Calculating the cost for the rhodium required (see Table 4.2), it becomes clear that the amount of investment needed is rather large for a privately owned device. One alternative is that the rhodium is leased by the owner of the fuel cell system. The yearly cost of the rhodium is then composed of the capital cost for the rhodium, the value of the losses and the cost of recycling the rhodium. For this evaluation, a loss of 4% of the catalyst mass and a recycling cost of 650 $/kg of noble metal were assumed [90]. For evaluation of the capital costs, two cases with an interest rate of 7%/a and of 15%/a respectively were used. A total lifetime of 10 years was assumed.

Table H.1 shows the rhodium price for 2006. In the recent past however, rhodium prices experienced a steady price increase. In 2001, a troy ounce (ca. 31.1 g) traded for $1600, in 2006 for $4555 and in 2008 for $6563 (all yearly averages [91]). To capture this significant increase in price, Table 4.2 shows the rhodium cost evaluated at all of these three price points.

Table 4.1 (a) summarizes the performance data and the cost available for the SulfaTrap^{TM} adsorbent developed by TDA research (USA)[92, 93]. The performance and the cost data is based on one field test over 2700 h in the Pittsburgh (USA) area with a sulfur loading of $Y_S = 11$ mg/m³. TDA Research used natural gas of similar composition to the gas used in this work. The compositions are given in Table 4.1 (b).

Apart from the adsorbent material cost, there is cost involved in packaging the adsorbent material in a suitable cartridge. This cost is estimated at

Table 4.1: (a) Performance data of the dedicated adsorbent developed by TDA Research (SulfaTrap^{TM}) and (b) Comparison of the natural gas compositions used (see [92, 93]).

(a)

Production cost	10–12 \$/lb ($\approx$ 22000–55000 \$/t)
Typical yearly cost for natural gas desulfurization, Y_S=11 mg/m^3	4.71 \$ for 1000 m^3 (NTP)

(b)

component	volume fraction y_i in %	
	this work (see Table 2.2)	TDA Research
CH_4	85.70	92.85
C_2H_6	5.20	3.30
C_3H_8	2.00	0.60
n-C_4H_{10}	0.30	0.13
i-C_4H_{10}	0.18	0.12
n-C_5H_{12}	0.01	0.30
CO_2	1.51	0.70
N_2	4.98	2.00

\$40 per cartridge, regardless of size. Furthermore, at an assumed hourly rate of 60 \$/h, and estimating 2 hour duration for the work (once per year) for the 1 and the 5 kW system, the labor cost is the largest contribution to the total yearly cost. In Table 4.2 the yearly cost estimated by adding these contributions to the adsorbent cost is included.

For the 250 kW system, it is clear that the rhodium cost is very high. Consequentially, even a leasing scheme is still be more costly than an adsorbent based option. For a 5 kW system, the situation is not much different, and though the difference between the two process options is smaller, the adsorbent based option is still more economical compared to the sulfur-tolerant processing. Only for small systems (1 kW) the yearly leasing cost for the rhodium is lower than the cost incurred by replacing the adsorbent. For these systems, the sulfur tolerant processing option would be more economical than the adsorbent based option.

Table 4.2: Required catalyst mass and associated lease cost as well as adsorbent cost. 10 years lifetime are assumed.

		$P_{el.}$ in kW		
		1	5	250
H_2 flow $^V\Phi_{H_2}$ (NTP)	ml/s	202	1030	51 856
NG flow $^V\Phi_{NG}$ (NTP)	ml/s	82	410	20 508
Sulfur tolerant catalyst				
catalyst mass	kg	0.53	2.67	133.4
total rhodium cost				
at 1 600 \$/tr oz (= 51\$/g), 2001	\$	274	1372	68 614
at 4 555 \$/tr oz (= 146\$/g), 2006	\$	782	3906	195 332
at 6 563 \$/tr oz (= 211\$/tg), 2008	\$	1502	7514	375 656
yearly rhodium lease (at 7%/a capital cost, 4% loss, 650 \$/kg recovery cost)				
at 1 600 \$/tr oz, 2001	\$	55	275	13 747
at 4 555 \$/tr oz, 2006	\$	92	462	23 124
at 6 563 \$/tr oz, 2008	\$	118	590	29 500
yearly rhodium lease (at 15%/a capital cost, 4% loss, 650 \$/kg recovery cost)				
at 1 600 \$/tr oz, 2001	\$	77	385	19 236
at 4 555 \$/tr oz, 2006	\$	155	775	38 751
at 6 563 \$/tr oz, 2008	\$	208	1040	52 019
yearly rhodium lease (at 15%/a capital cost, 2% loss, 650 \$/kg recovery cost)				
at 1 600 \$/tr oz, 2001	\$	76	382	19 099
at 4 555 \$/tr oz, 2006	\$	153	767	38 360
at 6 563 \$/tr oz, 2008	\$	206	1029	51 456
yearly rhodium lease (at 15%/a capital cost, 4% loss, 325 \$/kg recovery cost)				
at 1 600 \$/tr oz, 2001	\$	60	298	14 901
at 4 555 \$/tr oz, 2006	\$	138	688	34 416
at 6 563 \$/tr oz, 2008	\$	191	954	47 684
traditional catalyst and adsorbent				
adsorbent cost, E_n (adsorbent material only)	\$/a	12	61	3 040
yearly cost, E_n (adsorbent cartridge, & service)	\$/a	172	221	–

Considering the sensitivity of the yearly rhodium lease to the capital cost charged, the above conclusions do not change when the rate is increased from 7%/a to 15%/a, see Table 4.2. Also, further analysis shows that the loss of material has only very little influence on the yearly rhodium lease, while the recovery cost is a larger contribution. Nevertheless, a unrealistically large decrease in recovery cost would be required to bring the rhodium lease cost to lower than the yearly service cost for the adsorbent for the 5 kW system.

Overall, it can thus be concluded that the adsorbent process option is likely more economical in most applications. Only for small systems, the sulfur tolerant processing can be an attractive option, particularly when considering especially high labor cost for maintenance, e.g. in remote locations.

Chapter 5.

Summary, Conclusions and Outlook

In the introduction, the scope of this work was divided in 3 successive tasks:

- fabricate and characterize a catalyst similar to the one described by Krumpelt et al.

- test and explore it's catalytic properties and sulfur tolerance in the steam reforming of natural gas.

- use the gathered data to judge the feasibility of a sulfur-tolerant processing scheme.

In the preceding chapters, the experimental work was described, and it is now possible to summarize and draw conclusions regarding these items on this list.

Fabrication and characterization of the catalyst Various ceria and cerium gadolinium containing catalysts were fabricated, all having no microporosity and a low specific surface area of 3–2 m²/g. Up to a gadolinium fraction of about $x_{Gd} = 0.3$, the dopant is integrated into the fluorite lattice.

Similar to pure ceria, the material is reducible at higher temperatures, making it interesting as oxygen storage compound. Compared to pure ceria, the incorporation of gadolinum ions in the crystal lattice reduces the total oxygen storage capacity (OSC) at 800 °C while the kinetics of the (remaining) oxygen storage is enhanced.

Using pulse experiments, it was shown that sulfur and hydrogen sulfide interact with the support in a similar way as oxygen and water do. For gadolinium doped ceria, the ceria - sulfur bond seems to be weaker than for pure ceria.

Catalytic properties and sulfur tolerance The freshly prepared catalyst deactivates under typical operating conditions for the steam reforming of natural gas (800 °C, $\zeta = 3$ and natural gas as feed). Based on TPR and HR-TEM experiments and by analogy with results published in literature regarding the SMSI effect, however, an activation procedure leading to a stable catalyst was developed. Further pulse experiments showed that the reducibility of the support results in the lattice oxygen participating in reactions with carbon, which results in an increased tolerance against coking and improves the stability of the catalyst.

Measurements on a series of 1 wt.-% Rh CGO catalysts with different gadolinium contents showed that there is no clear trend for the activation energy or the activity (measured as turnover frequency) with the gadolinium content.

Upon addition of sulfur to the feed a large reduction in activity of the catalyst is observed, both for the shift as well as for the reforming reaction. Nevertheless, it was shown that the catalyst retains some reforming activity for a sulfur compounds containing feed over at least 100 h. The negative effect of sulfur is reversible, a conclusion drawn from both continuous flow experiments as well as from pulse experiments. The latter further detailed that both the interaction of hydrogen sulfide with rhodium and of sulfur with the support is reversible. In addition to the tolerance towards sulfur components in the feed, the sulfur components are fully converted. H_2S is the main sulfurous reaction product.

The extend to which the catalyst retains activity despite sulfur containing compounds in the feed is non-linear, with very small amounts of sulfur containing compounds in the feed already suppressing the activity significantly. Given the reversibility of the adsorption, there is a large influence of the operating conditions on the retained activity. Generally, all factors leading to a high hydrogen partial pressure are beneficial for the catalyst's activity under sulfur containing feed.

Under typical steam reforming conditions (800 °C, $\zeta = 3$) and with a typical natural gas composition (see Table 2.2) and with a sulfur loading Y_S of 10-15 mg/m^3, an effective methane reaction constant $k_{CH_4} = 2\text{--}7\,ml/g{\cdot}s$ can be achieved. This corresponds to approximately 3-7% of the activity under sulfur free conditions.

The successful modelling of the catalysts behavior using an approach that takes the dependence of the reforming activity on the rhodium surface coverage by sulfur into account shows that the reforming activity is mainly due to the rhodium, even under sulfurous conditions. As shown earlier, the presence of gadolinium doped ceria as support creates the possibility for the lattice oxygen to participate in the steam reforming reaction. The modeling results suggest that this reduces the ensemble size required for the steam reforming reaction, making the impact of sulfur adsorption on rhodium much less severe. Herein could lie one of the reasons for the observed sulfur tolerance.

Economic feasibility   A brief economic comparison showed that an alternative process option using a dedicated adsorbent for sulfur removal upstream the steam reforming reactor is likely more economical, especially for large systems (250 kW electrical output). Nevertheless, the sulfur-tolerant option can be attractive for small systems (1 kW electrical output) and in remote locations, where the cost of replacing the adsorbent is high.

Regarding the initially formulated scope of this study, it is concluded that the sulfur-tolerant processing scheme is, although technically possible, only feasible as a niche application and is not suitable for decentralized combined heat and power generation in general.

Outlook   Several improvements can be envisioned in regards to the catalyst preparation. First, the catalysts prepared here have rather low specific surface area. Although high surface area ceria is likely to sinter under operating conditions due to the high operating temperature and the high water partial pressure, it would be worthwhile to try to increase the specific surface area of the support, e.g. by alternative preparation techniques such as the citrate method (e.g. [59]). With various sintering experiments under model operating conditions the material with the highest stable specific surface area could be determined. Combined with this topic is the small particle size of the material, which led to a large pressure drop in the experiments presented here. It would be desirable to increase the particle size to achieve a lower pressure drop in the reactor. If the same activity is to be retained, this will however require a porous material with internal surface area and thus will give rise to mass-transfer limitations.

A second area of improvement is the amount of rhodium metal in the catalyst. In this work, a rhodium loading ω_{Rh} of 1 wt.-% was used - mainly for practical reasons when characterizing the catalyst. Synthesizing a range of catalysts with lower rhodium content therefore should be done. This goes hand in hand with increasing the catalyst's dispersion. Care should be taken when testing the catalyst's activity though, because similar to the support surface area, sintering of the noble metal could occur under operating conditions. Thus, during the experiments, it has to be made sure that the measured activity can be sustained over longer periods of time. Both, lower rhodium loading and higher dispersion could be achieved through alternative preparation methods, such as the incipient wetness method. When choosing the rhodium precursor, rhodium chloride should be avoided, because it was proven that this leads to catalysts with a modified rhodium/ceria interface [59].

Given the results presented in section 3.1.2, it is possible that other noble metals supported on gadolinium doped ceria, such as platinum or ruthenium show stable reforming activity when subjected to redox cycles. The topic of catalyst stability for these catalysts should therefore be revised. Equally, other elements than gadolinium as dopants can be tried. Preference should be give to zirconium doped ceria, as this increases the oxygen storage capacity (OSC) compared to pure ceria. This is in contrast to gadolinium doped ceria, and pulse experiments similar to the ones here as well as steam reforming experiments with sulfurous feed would clarify the influence of the dopant on the interaction of the catalyst with sulfur.

Ideally, the final catalyst form will likely be a monolith with a noble metal/doped ceria wash coat, similar to a three way catalyst. This form of catalyst, however will be difficult to characterize and may be subject to transport limitations. Such development therefore should only be undertaken with the goal of designing a practical reforming device with a given catalyst, not with the goal of improving the catalyst's intrinsic properties. To lay a better foundation for designing such a reactor, more experiments exploring the kinetics of the natural gas reforming without and with sulfur in the feed need to be carried out. With a view on more fundamental insights, more experiments allowing for more accurate parameter estimation regarding the influence of sulfur on the steam reforming reaction are desirable.

Nomenclature

A	area, general
a	unit-cell parameter
A_R	reactor cross-sectional area
b	parameter used for relating the activity factor φ to the rhodium surface coverage, see equation 3.28
c	parameter used for relating the activity factor φ to the rhodium surface coverage, see equation 3.28
$c_{i(,z)}$	concentration of species i (at location z); $c_{i,z} = \frac{n_i}{V}$
d_p	particle diameter
d_R	reactor inner diameter
a	distance between two crystallographically identical planes
$D_{ii'}$	binary diffusion coefficient for species i and i′
d_A	particle agglomerate diameter
D_{Rh}	Rhodium dispersion; $D_{Rh} = \frac{n_{Rh,surface}}{n_{Rh}}$
E_A	activation energy. here, only the methane steam reforming reaction is considered
e^-	charge of one electron
f	friction factor
F	force
g	gravitational constant

$H_{I(,i)}$	lower heating value (of species or mixture i)
$H_{S(,i)}$	higher heating value (of species or mixture i)
k_j	reaction rate constant for reaction k
k'_{CH_4}	reaction rate constant for steam reforming reaction, using the simplified method (see Equation 2.18)
k_{CH_4}	modified reaction rate constant for steam reforming reaction; $k'_{CH_4} = \dfrac{k_{CH_4}}{\rho_{bed}}$
$K_{ads,i}$	surface adsorption constant of species i
$K_{p(,eq)}$	reaction equilibrium constant (in equilibrium), see Equation 3.16
m	mass, general
m_i	mass of species i
m_{cat}	catalyst mass (in reactor)
M_i	molar mass of species i
OEC	dynamic oxygen exchange capacity; see Appendix C
OSC	total oxygen storage capacity; see Appendix C
p	pressure, general; $p = \frac{F}{A}$
$\bar{p}$	average pressure in the reactor; see Equation 2.14
$p_{R(,z)}$	pressure in reactor (at location z)
Δp	pressure drop in the reactor; $\Delta p = p_{R,in} - p_{R,out}$
Q_i	electrical charge of ion i
r_j	reaction rate of reaction j; $r_j = \frac{1}{V_R}\frac{1}{v_{i,j}}\frac{dn_i}{dt}$
Re_p	Reynolds number, related to the particle; $Re_p = \frac{u_0 d_p \rho}{\mu}$

RSS	residual sum of squares, see Equation 3.19
S_{BET}	specific surface area, determined with BET method; see chapter 2.2.1
$S_{C2=,z}$	ethene selectivity; $S_{C2=,z} = \dfrac{{}^{N}\Phi_{C_2H_4,z}}{\sum_i Y_{C,i}({}^{N}\Phi_{i,in}-{}^{N}\Phi_{i,z})}$ for i= CH_4, C_2H_6, C_3H_8, C_4H_{10}, C_5H_{12}, CO_2
$S_{C3=,z}$	propene selectivity; $S_{C3=,z} = \dfrac{{}^{N}\Phi_{C_3H_6,z}}{\sum_i Y_{C,i}({}^{N}\Phi_{i,in}-{}^{N}\Phi_{i,z})}$ for i= CH_4, C_2H_6, C_3H_8, C_4H_{10}, C_5H_{12}, CO_2
$S_{CO,z}$	CO selectivity; $S_{CO,z} = \dfrac{{}^{N}\Phi_{CO,z}}{\sum_i Y_{C,i}({}^{N}\Phi_{i,in}-{}^{N}\Phi_{i,z})}$ for i= CH_4, C_2H_6, C_3H_8, C_4H_{10}, C_5H_{12}, CO_2
$S_{CO_2,z}$	CO_2 selectivity; $S_{CO_2,z} = \dfrac{{}^{N}\Phi_{CO_2,z}}{\sum_i Y_{C,i}({}^{N}\Phi_{i,in}-{}^{N}\Phi_{i,z})}$ for i= CH_4, C_2H_6, C_3H_8, C_4H_{10}, C_5H_{12}, CO_2
SR	standardized residual, see Equation 3.18
T_{calc}	calcination temperature
T_R	reactor temperature
T	temperature, general
t	time, general
T_{red}	reduction temperature
TOF_{CH_4}	turnover frequency of the methane reforming reaction over rhodium; $TOF_{CH_4} = k_{CH_4}(T,P)\dfrac{y_{CH_4}}{V_{m,}(T,p)}\dfrac{M_{Rh}}{Y_{Rh}D_{Rh}}$
$u_{0(,z)}$	superficial velocity (at location z); $u_0,z = \dfrac{{}^{V}\Phi_z}{A_R}$
$\vec{u}$	velocity vector
u_z	velocity in direction z
V	gas volume, general

$V_m(T,p)$ molar volume of an ideal gas at temperature T and pressure p.

$V_{m,NTP}$ molar volume of an ideal gas at norm conditions. $V_{m,NTP}$=22.4 l/mol

V_R (empty) reactor volume

v_i volume of species i in a mixture

$x_{i(,z)}$ mol fraction of species i (at location z); $x_i = \frac{n_i}{\sum_{all\,i} n_i}$

$X_{i(,z)}$ conversion of species i (at location z); $X_{CH_4} = \frac{{}^N\Phi_{CH_4,in} - {}^N\Phi_{CH_4}}{{}^N\Phi_{CH_4}}$

Y_i loading of species i; $Y_i = \frac{m_i}{V}$

Y_S (total) sulfur loading

$Y_{S,base}$ sulfur base loading, i.e. natural gas without odorant

$y_{i,z}$ volume fraction of species i (at location z); $y_i = \frac{v_i}{V}$

z charge of an ion, in multiples of one electron; $z=Q_i/e^-$

α_j exponent in power-law rate equation for reaction j, see Equations 3.14, 3.15 and 3.16

β_j exponent in power-law rate equation for reaction j, see Equations 3.14, 3.15 and 3.16

δ unit vector; $\delta = ((1\ 0\ 0)(0\ 1\ 0)(0\ 0\ 1))$

ε catalyst bed void fraction; $\varepsilon = \frac{V}{V_R}$

${}^N\Phi_{C(,z)}$ molar flow of carbon (at location z); ${}^N\Phi_{C(,z)} = \frac{d\sum_{all\,i} \Xi_{C,i} * n_i}{dt}$

${}^N\Phi_{i(,z)}$ molar flow of species i (at location z), ${}^N\Phi_i = \frac{dn_i}{dt}$

${}^V\Phi_z$ total volume flow at location z, ${}^V\Phi = \frac{dV}{dt}$

Γ shear stress tensor

$\nu_{i,j}$ stoichiometric coefficient of species i in reaction j

λ	wavelength
μ	kinematic viscosity; $\mu = \frac{v}{\rho}$
v	dynamic viscosity; $v = \frac{\partial}{\partial z}\frac{\vec{u}}{\Gamma}$
ω_{Rh}	mass fraction of rhodium metal in the catalyst; $\omega_{Rh} = \frac{m_{Rh}}{m_{cat}}$
Ψ	fractional molar increase; see Equations 2.16 and 2.17
φ	activity factor, see Equation 3.26
ρ	density; $\rho = \frac{m}{V}$
ρ_{bed}	catalyst bed density; $\rho_{bed} = \frac{m_{cat}}{V_R}$
Θ	diffraction angle
$\theta_{Rh,S}$	surface coverage of the rhodium with sulfur; $\theta_{Rh,S} = \frac{n_{S,surface}}{n_{Rh,surface}}$, i.e. assuming an adsorption stoichiometry of 1 sulfur atom per Rh atom
τ	reactor residence time; $\tau = \frac{V_R}{V_{\Phi_{in}}}$
τ_{mod}	modified reactor residence time; $\tau_{mod} = \rho_{bed}\tau$
$\Xi_{C,i}$	amount of carbon atoms in one mol of species i
ζ	steam to carbon ratio; $\zeta = \frac{{}^N\Phi_{H_2O}}{{}^N\Phi_C}$

Bibliography

[1] C. Bartholomew, Mechanisms of catalyst deactivation, Applied Catalysis A: General 212 (2001) 17–60.

[2] J. Meusinger, E. Riensche, U. Stimming, Reforming of natural gas in solid oxide fuel cell systems, Journal of Power Sources 71 (1-2) (1998) 315–320.

[3] M. Adelt, G. Dubielzig, M. Brune, F. Gröschl, D. Hentze, R. Reimert, ENBA - A project formulating basic principles for the standardization of the fuel cell technology, GWF, Gas - Erdgas 146 (2) (2005) 104–111.

[4] Technische Regel Arbeitsblatt G-260, Gasbeschaffenheit, DVGW - Deutsche Vereinigung des Gas- und Wasserfachs e.V. (2000).

[5] Technische Regel Arbeitsblatt G-260, Gasbeschaffenheit, DVGW - Deutsche Vereinigung des Gas- und Wasserfachs e.V. (2008).

[6] Technische Regel Arbeitsblatt G-280-1, Gasodorierung, DVGW - Deutsche Vereinigung des Gas- und Wasserfachs e.V. (2003).

[7] J. Rostrup-Nielsen, Steam reforming of hydrocarbons. A historical perspective, in: Natural Gas Conversion VII, Vol. 147 of Studies in Surface Science and Catalysis, 2004, pp. 121–126.

[8] J. Rostrup-Nielsen, Catalytic Steam Reforming, in: J. Andersen, M. Boudard (Eds.), Catalysis - Science and Technology, Vol. 5, Springer Verlag, Berlin, Heidelberg, 1984, Ch. 1, pp. 3–117.

[9] T. Rostrup-Nielsen, Manufacture of hydrogen, Catalysis Today 106 (2005) 293–296.

[10] J. Rostrup-Nielsen, R. Nielsen, Fuels and energy for the future: The role of catalysis, Catalysis Reviews-Science And Engineering 46 (2004) 247–270.

[11] K. Aasberg-Petersen, J. Hansen, T. Christensen, I. Dybkjaer, P. Christensen, C. Nielsen, S. Madsen, J. Rostrup-Nielsen, Technologies for large-scale gas conversion, Applied Catalysis A: General 221 (2001) 379–387.

[12] H.-J. Renner, R. Marschner, Gasproduction, in: Ullmann's Encyclopedia of Industrial Chemistry, 7th Edition, Wiley-VCH, 2003.

[13] P. Häussinger, R. Lohmüller, A. Watson, Hydrogen, in: Ullmann's Encyclopedia of Industrial Chemistry, 7th Edition, Wiley-VCH, 2003.

[14] V. M. Janardhanan, O. Deutschmann, CFD analysis of a solid oxide fuel cell with internal reforming: Coupled interactions of transport, heterogeneous catalysis and electrochemical processes, Journal of Power Sources 162 (2) (2006) 1192–1202.

[15] A. Faghri, Z. Guo, Challenges and opportunities of thermal management issues related to fuel cell technology and modeling, International Journal of Heat and Mass Transfer 48 (2005) 3891–3920.

[16] G. Kolios, A. Gritsch, A. Morillo, U. Tuttlies, J. Bernnat, F. Opferkuch, G. Eigenberger, Heat-integrated reactor concepts for catalytic reforming and automotive exhaust purification, Applied Catalysis B: Environmental 70 (1-4) (2007) 16–30.

[17] B. Glockler, G. Kolios, G. Eigenberger, Analysis of a novel reverse-flow reactor concept for autothermal methane steam reforming, Chemical Engineering Science 58 (3-6) (2003) 593–601.

[18] H. Bengaard, J. Norskov, J. Sehested, B. Clausen, L. Nielsen, A. Molenbroek, J. Rostrup-Nielsen, Steam reforming and graphite formation on Ni catalysts, Journal Of Catalysis 209 (2002) 365–384.

[19] J. Sehested, J. A. P. Gelten, S. Helveg, Sintering of nickel catalysts: Effects of time, atmosphere, temperature, nickel-carrier interactions, and dopants, Applied Catalysis A: General 309 (2) (2006) 237–246.

[20] J. Sehested, Four challenges for nickel steam-reforming catalysts, Catalysis Today 111 (1-2) (2006) 103–110.

[21] C. Bartholomew, P. Agrawal, J. Katzer, Sulfur poisoning of metals, Advances In Catalysis 31 (1982) 135–242.

[22] C. M. Friend, D. A. Chen, Fundamental studies of hydrodesulfurization by metal surfaces, Polyhedron 16 (18) (1997) 3165–3175.

[23] P. D. N. Svoronos, T. J. Bruno, Carbonyl sulfide: A review of its chemistry and properties, Industrial & Engineering Chemistry Research 41 (22) (2002) 5321–5336.

[24] J. Rostrup-Nielsen, Some principles relating to regeneration of sulfur-poisoned nickel catalyst, Journal Of Catalysis 21 (1971) 171–178.

[25] L. J. Christiansen, S. L. Andersen, Transient profiles in sulphur poisoning of steam reformers, Chemical Engineering Science 35 (1-2) (1980) 314–321.

[26] H. Roh, K. Jun, W. Dong, S. Park, Y. Baek, Highly stable Ni catalyst supported on Ce-ZrO_2 for oxy-steam reforming of methane, Catalysis Letters 74 (2001) 31–36.

[27] H. Roh, H. Potdar, K. Jun, J. Kim, Y. Oh, Carbon dioxide reforming of methane over Ni incorporated into Ce-ZrO_2 catalysts, Applied Catalysis A: General 276 (2004) 231–239.

[28] F. Passos, E. de Oliveira, L. Mattos, F. Noronha, Partial oxidation of methane to synthesis gas on $Pt/Ce_xZr_{1-x}O_2$ catalysts: the effect of the support reducibility and of the metal dispersion on the stability of the catalysts, Catalysis Today 101 (2005) 23–30.

[29] R. Ahlborn, F. Baumann, S. Wieland, EP 1 157 968 A1: Verfahren zur autothermen Dampfreformierung von Kohlenwasserstoffen, to: OMG AG, Hanau, Germany (2001).

[30] M. Krumpelt, S. Ahmed, R. Kumar, R. Doshi, US Patent 6,110,861, Partial Oxidation Catalyst, to: University of Chicago, Chicago, USA (2002).

[31] M. Krumpelt, T. Krause, J. Carter, J. Kopasz, S. Ahmed, Fuel processing for fuel cell systems in transportation and portable power applications, Catalysis Today 77 (2002) 3–16.

[32] M. Mogensen, N. Sammes, G. Tompsett, Physical, chemical and electrochemical properties of pure and doped ceria, Solid State Ionics 129 (2000) 63–94.

[33] A. Weber, E. Ivers-Tiffée, Materials and concepts for solid oxide fuel cells (SOFCs) in stationary and mobile applications, Journal Of Power Sources 127 (2004) 273–283.

[34] R. Purohit, B. Sharma, K. Pillai, A. Tyagi, Ultrafine ceria powders via glycine-nitrate combustion, Materials Research Bulletin 36 (2001) 2711–2721.

[35] U. Hennings, R. Reimert, Noble metal catalysts supported on gadolinium doped ceria used for natural gas reforming in fuel cell applications, Applied Catalysis B: Environmental 70 (2007) 498–508.

[36] D. Fino, N. Russo, E. Cauda, G. Saracco, V. Specchia, La-Li-Cr perovskite catalysts for diesel particulate combustion, Catalysis Today 114 (2006) 31–39.

[37] J. Laugier, A. Fihol, Eracel (http://pcb4122.univ-lemans.fr/also.html, retrieved 13.03.208).

[38] K. Sing, J. Rouquerol, Surface Area and Porosity, in: G. Ertl, H. Knörzinger, J. Weitkamp (Eds.), Handbook of Heterogenous Catalysis, Vol. 2, Wiley-VCH, Weinheim, 1997, Ch. 3, pp. 427–438.

[39] S. Brunauer, P. H. Emmett, E. Teller, Adsorption of gases in multimolecular layers, Journal of the American Chemical Society 60 (2) (1938) 309–319.

[40] J. Gatica, R. Baker, P. Fornasiero, S. Bernal, G. Blanco, J. Kaspar, Rhodium dispersion in a $Rh/Ce_{0.68}Zr_{0.32}O_2$ catalyst investigated by HRTEM and H-2 chemisorption, Journal Of Physical Chemistry B 104 (2000) 4667–4672.

[41] G. Bergeret, P. Gallezot, Particle Size and Dispersion Measurements, in: G. Ertl, H. Knörzinger, J. Weitkamp (Eds.), Handbook of Heterogenous Catalysis, Vol. 2, Wiley-VCH, Weinheim, 1997, Ch. 3, pp. 439–464.

[42] W. Stark, M. Maciejewski, L. Madler, S. Pratsinis, A. Baiker, Flamemade nanocrystalline ceria/zirconia: structural properties and dynamic oxygen exchange capacity, Journal Of Catalysis 220 (2003) 35–43.

[43] W. Stark, J. Grunwaldt, M. Maciejewski, S. Pratsinis, A. Baiker, Flame-made Pt/ceria/zirconia for low-temperature oxygen exchange, Chemistry of Materials 17 (2005) 3352–3358.

[44] Bird, Steward, Lightfoot, Transport Phenomena, John Wiley & Sons, 1960.

[45] R. H. Perry, D. W. Green (Eds.), Perry's Chemical Engineers' Handbook, 7th Edition, McGraw-Hill, 1997.

[46] G. F. Froment, K. Bischoff, Chemical Reactor Analysis and Design, 2nd Edition, John Wiley & Sons, 1990.

[47] P. N. Brown, A. Hindmarsh, L. Petzold, Using Krylov methods in the soluiton of large-scale differential-algebraic systems, SIAM J. Sci. Comput. 15 (1994) 1467–1488.

[48] R. R.C., P. J.M., B. Pauling, The properties of Gases and Liquids - their estimation and correlation, 4th Edition, McGraw-Hill, New York, 1987.

[49] O. Levenspiel, Chemical Reaction Engineering, 2nd Edition, John Wiley & Sons, 1972.

[50] R. Gorte, T. Luo, SO_2 Poisoning of Ceria-supported, Metal Catalysts, in: A. Trovarelli (Ed.), Catalysis by Ceria and Related Materials, Vol. 2 of Catalytic Sciences Series, Imperial College Press, London, 2002, Ch. 4, pp. 377–389.

[51] K. Yi, E. Podlaha, D. Harrison, Ceria-zirconia high-temperature desulfurization sorbents, Industrial & Engineering Chemistry Research 44 (2005) 7086–7091.

[52] U. Hennings, R. Reimert, Stability of rhodium catalysts supported on gadolinium doped ceria under steam reforming conditions, Applied Catalysis A: General 337 (1) (2008) 1–9.

[53] U. Hennings, R. Reimert, Investigation of the structure and the redox behavior of gadolinium doped ceria to select a suitable composition for use as catalyst support in the steam reforming of natural gas, Applied Catalysis A: General 325 (2007) 41–49.

[54] S. Dikmen, P. Shuk, M. Greenblatt, H. Gocmez, Hydrothermal synthesis and properties of $Ce_{1-x}Gd_xO_{2-\delta}$ solid solutions, Solid State Sciences 4 (2002) 585–590.

[55] S. Hong, A. Virkar, Lattice-Parameters and Densities of Rare-Earth-Oxide doped Ceria Electrolytes, Journal Of The American Ceramic Society 78 (1995) 433–439.

[56] D. Kim, Lattice-Parameters, Ionic Conductivities, And Solubility Limits In Fluorite-Structure $Hf^{4+}O_2$, $Zr^{4+}O_2$, $Ce^{4+}O_2$, $Th^{4+}O_2$, $V^{4+}O_2$ Oxide Solid-Solutions, Journal Of The American Ceramic Society 72 (1989) 1415–1421.

[57] E. Ramirez-Cabrera, A. Atkinson, D. Chadwick, Reactivity of ceria, Gd- and Nb-doped ceria to methane, Applied Catalysis B: Environmental 36 (2002) 193–206.

[58] E. Aneggi, J. Llorca, M. Boaro, A. Trovarelli, Surface-structure sensitivity of CO oxidation over polycrystalline ceria powders, Journal Of Catalysis 234 (2005) 88–95.

[59] S. Bernal, J. J. Calvino, J. M. Gatica, C. L. Cartes, J. M. Pintado, Chemical and nanostructural aspects of the preparation and characterization of ceria and ceria-based mixed oxide supported catalysts, in: A. Trovarelli (Ed.), Catalysis by Ceria and Related Materials, Vol. 2 of Catalytic Sciences Series, Imperial College Press, London, 2002, Ch. 4, pp. 85–168.

[60] A. Trovarelli, Structural properties and nonstoichiometric behavior of CeO_2, in: A. Trovarelli (Ed.), Catalysis by Ceria and Related Materials, Vol. 2 of Catalytic Sciences Series, Imperial College Press, London, 2002, Ch. 4, pp. 15–50.

[61] F. A. Kröger, The Chemistry of Imperfect Crystals, 2nd Edition, Vol. 2, North-Holland Publishing Company, Amsterdam, Holland, 1974.

[62] G. Balducci, M. Islam, J. Kaspar, P. Fornasiero, M. Graziani, Reduction process in CeO_2-MO and CeO_2-M_2O_3 mixed oxides: A computer simulation study, Chemistry of Materials 15 (2003) 3781–3785.

[63] D. Y. Wang, D. S. Park, J. Griffith, A. S. Nowick, Oxygen-ion conductivity and defect interactions in yttria-doped ceria, Solid State Ionics 2 (2) (1981) 95–105.

[64] S. Bernal, J. Calvino, M. Cauqui, J. Gatica, C. Larese, J. Omil, J. Pintado, Some recent results on metal/support interaction effects in NM/CeO_2 (NM : noble metal) catalysts, Catalysis Today 50 (1999) 175–206.

[65] J. Gatica, R. Baker, P. Fornasiero, S. Bernal, J. Kaspar, Characterization of the metal phase in $NM/Ce_{0.68}Zr_{0.32}O_2$ (NM : Pt and Pd) catalysts by hydrogen chemisorption and HRTEM microscopy: A comparative study, Journal Of Physical Chemistry B 105 (2001) 1191–1199.

[66] S. Penner, D. Wang, R. Podloucky, R. Schlögl, K. Hayek, Rh and Pt nanoparticles supported by CeO_2: Metal-support interaction upon high-temperature reduction observed by electron microscopy, Physical Chemistry Chemical Physics 6 (2004) 5244–5249.

[67] S. Tauster, Strong metal-support interactions, Accounts Of Chemical Research 20 (1987) 389–394.

[68] K. Hedden, A. Wilhelm, Catalytic effects of inorganic substances on reactivity and ignition temperature of solid fuels, Ger. Chem. Eng. (3) (1980) 142–147.

[69] H. Wang, E. Ruckenstein, Carbon dioxide reforming of methane to synthesis gas over supported rhodium catalysts: the effect of support, Applied Catalysis A: General 204 (2000) 143–152.

[70] M. Wisniewski, A. Boreave, P. Gelin, Catalytic CO_2 reforming of methane over $Ir/Ce_{0.9}Gd_{0.1}O_{2-x}$, Catalysis Communications 6 (9) (2005) 596–600.

[71] J. Wei, E. Iglesia, Structural requirements and reaction pathways in methane activation and chemical conversion catalyzed by rhodium, Journal Of Catalysis 225 (2004) 116–127.

[72] Z. Liu, P. Hu, General rules for predicting where a catalytic reaction should occur on metal surfaces: A density functional theory study of C-H and C-O bond breaking/making on flat, stepped, and kinked metal surfaces, Journal Of The American Chemical Society 125 (2003) 1958–1967.

[73] J. R. Rostrup-Nielsen, Coking on nickel catalysts for steam reforming of hydrocarbons, Journal of Catalysis 33 (2) (1974) 184–201.

[74] H. Karjalainen, U. Lassi, K. Rahkamaa-Tolonen, V. Kroger, R. Keiski, Thermodynamic equilibrium calculations of sulfur poisoning in Ce-O-S and La-O-S systems, Catalysis Today 100 (2005) 291–295.

[75] R. J. Gorte, S. Zhao, Studies of the water-gas-shift reaction with ceria-supported precious metals, Catalysis Today 104 (1) (2005) 18–24.

[76] J. Wei, E. Iglesia, Isotopic and kinetic assessment of the mechanism of methane reforming and decomposition reactions on supported iridium catalysts, Physical Chemistry Chemical Physics 6 (2004) 3754–3759.

[77] J. Wei, E. Iglesia, Reaction pathways and site requirements for the activation and chemical conversion of methane on ru-based catalysts, Journal Of Physical Chemistry B 108 (2004) 7253–7262.

[78] J. Wei, E. Iglesia, Isotopic and kinetic assessment of the mechanism of reactions of CH_4 with CO_2 or H_2O to form synthesis gas and carbon on nickel catalysts, Journal Of Catalysis 224 (2004) 370–383.

[79] B. Schädel, O. Deutschmann, Steam reforming of natural gas on nobel-metal based catalysts: Predictive modeling, Studies in Surface Science and Catalysis 167 (2007) 207–212.

[80] J. J. Krummenacher, K. N. West, L. D. Schmidt, Catalytic partial oxidation of higher hydrocarbons at millisecond contact times: Decane, hexadecane, and diesel fuel, Journal of Catalysis 215 (2) (2003) 332–343.

[81] J. J. Krummenacher, L. D. Schmidt, High yields of olefins and hydrogen from decane in short contact time reactors: Rhodium versus platinum, Journal of Catalysis 222 (2) (2004) 429–438.

[82] B. Vogel, Experimentelle und analytische Untersuchungen zur Wasserstofferzeugung für Membran-Brennstoffzellen, Ph.D. thesis, Universität Karlsruhe(TH) (2004).

[83] J. M. Thomas, W. J. Thomas, Principle and Practice of Heterogeneous Catalysis, Wiley-VCH, 1996.

[84] A. Donazzi, A. Beretta, G. Groppi, P. Forzatti, Catalytic partial oxidation of methane over a 4% Rh α-Al$_2$O$_3$ catalyst: Part I: Kinetic study in annular reactor, Journal of Catalysis 255 (2) (2008) 241–258.

[85] J. R. Rostrup-Nielsen, Sulfur-passivated nickel catalysts for carbon-free steam reforming of methane, Journal of Catalysis 85 (1) (1984) 31–43.

[86] J. R. Rostrup-Nielsen, Chemisorption of hydrogen sulfide on a supported nickel catalyst, Journal of Catalysis 11 (3) (1968) 220–227.

[87] H. A. Yoon, M. Salmerona, G. A. Somorjai, An LEED and STM study of the structures formed by sulfur on the Rh(111) surface, Surface Science 395 (2-3) (1998) 268–279.

[88] C. J. Zhang, P. Hu, M. H. Lee, A density functional theory study on the interaction between chemisorbed CO and S on Rh(111), Surface Science 432 (3) (1999) 305–315.

[89] J. G. McCarty, H. Wise, Thermodynamics of sulfur chemisorption on metals. i. alumina-supported nickel, J. Chem. Phys. 72 (12) (1980) 6332–6337.

[90] Evonik, Catalyst Cost Calculation Tool (http://www.degussa-catalysts.com/ec2/catalysts/calculation_tool.htm), accessed 18.01.2009.

[91] Johnson Matthey, Platinum Today (http://www.platinum.matthey.com/), accessed 25.06.2008.

[92] G. O. Alptekin, S. DeVoss, M. Dubovik, J. Monroe, R. Amaltitano, G. Isarelson, Regenerable Sorbent for Natural Gas Desulfurization, Journal of Materials Engineering and Performance 15 (433-438).

[93] G. O. Alptekin, Sorbents for Desulfurization of Natural Gas, LPG and Transportation Fuels, Presentation at the 6th Annual SECA Workshop, Pacific Grove, CA, USA (April 2004).

[94] R. J. Berger, F. Kapteijn, J. A. Moulijn, Overview of requirements for measurement of intrinsic kinetics in the G-S and L-S fixed-bed reactor, Tech. rep., Eurokin (1998).

[95] J. A. Moulijn, A. Tarfaoui, F. Kapteijn, General aspects of catalyst testing, Catalysis Today 11 (1991) 1–12.

[96] J. Pérez-Ramírez, R. J. Berger, G. Mul, F. Kapteijn, J. A. Moulijn, The six-flow reactor technology: A review on fast catalyst screening and kinetic studies, Catalysis Today 60 (1-2) (2000) 93–109.

[97] A. Slack, G. James, Ammonia, Vol. 1, Marcel Dekker Inc., New York, 1973.

[98] A. Ozaki, K. Akia, in: J. Anderson, M. Boudard (Eds.), Catalysis - Science and Technology, Vol. 1, Springer, 1981, p. 87.

[99] S. Dahl, A. Logadottir, C. J. H. Jacobsen, J. K. Norskov, Electronic factors in catalysis: the volcano curve and the effect of promotion in catalytic ammonia synthesis, Applied Catalysis A: General 222 (1-2) (2001) 19–29.

[100] E. L. Bray, Minerals yearbook - aluminium, Tech. rep., US Geological Survey (2006).

[101] D. A. Kramer, Minerals yearbook - magnesium compounds, Tech. rep., US Geological Survey (2006).

[102] P. H. Kuck, Minerals yearbook - nickel, Tech. rep., US Geological Survey (2006).

[103] M. W. George, Minerals yearbook - platinum-group metals, Tech. rep., US Geological Survey (2006).

[104] J. B. Hedrick, Minerals yearbook - rare earths, Tech. rep., US Geological Survey (2006).

[105] A. C. Tolcin, Minerals yearbook - zinc, Tech. rep., US Geological Survey (2006).

Appendix A.

Typical catalyst preparation

During the synthesis of $Ce_{0.8}Gd_{0.2}O_{1.9-\delta}$ the following reaction takes place:

$$0.8Ce(NO_3)_3 \cdot 6H_2O + 0.2Gd(NO_3)_3 \cdot 6H_2O + \frac{15}{9}NH_2CH_2COOH + 0.15O_2$$

$$\rightarrow Ce_{0.8}Gd_{0.2}O_{1.9-\delta} + \frac{7}{3}N_2 + \frac{10}{3}CO_2 + \frac{61}{6}H_2O$$

Given the above stoichiometry, and adding in the rhodium precursor the following amounts of precursors are needed to make 1.5 g of 1 wt.-% $Rh/Ce_{0.8}Gd_{0.2}O_{1.9-\delta}$ material. All used chemicals were supplied by Alfa Aesar, Karlsruhe, Germany. The required calculation was implemented in an excel sheet that allowed for quick calculation of the required amounts upon entry of the desired carrier composition and metal loading.

	M_i in g/mol	amount in mmol	mass in mg
$Ce(NO_3)_3 \cdot 6\,H_2O$	434.22	6.908	3000
$Gd(NO_3)_3 \cdot 6\,H_2O$	451.26	1.727	779
$Rh(NO_3)_3 \cdot 2\,H_2O$	433.01	0.077	30
glycine	75.07	14.393	1081

One attempt to measure the temperature inside the pressure vessel during the synthesis was undertaken. As can be seen from Figure A.1 it can be determined that ignition takes place at 200 °C, and that temperatures in excess of 500 °C are reached. Purohit calculated adiabatic flame temperatures above 1200 °C [34], and while the sluggish response from the thermocouple employed certainly contributes to the much lower termperature measured, it is also clear that such high temperatures are only prevalent during a very short time.

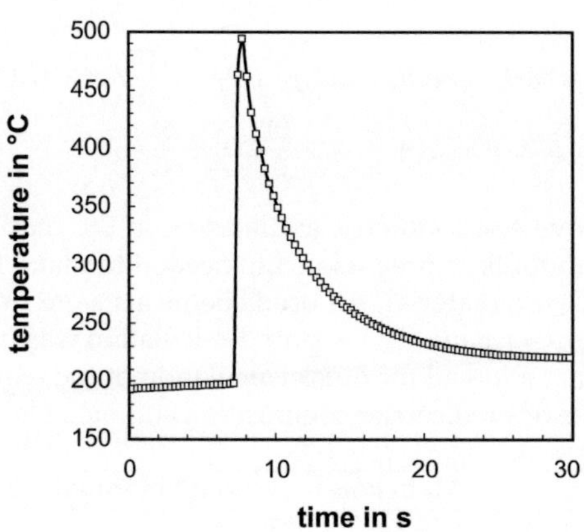

Figure A.1: Temperature progression during the catalyst synthesis

Appendix B.

Bulk elemental composition

The molar gadolinium fractions measured by ICP-OES reflect fairly well the values adjusted during synthesis (see Table B.1), indicating that the chemical composition of the materials is fully controlled by the composition of the synthesis solution.

Table B.1: Elemental analysis of the samples

Material	x_{Gd}
$Ce_{0.95}Gd_{0.05}O_{1.975-\delta}$	0.054
$Ce_{0.9}Gd_{0.1}O_{1.95-\delta}$	0.104
$Ce_{0.8}Gd_{0.2}O_{1.9-\delta}$	0.207
$Ce_{0.7}Gd_{0.3}O_{1.85-\delta}$	0.294
$Ce_{0.6}Gd_{0.6}O_{1.80-\delta}$	0.401
$Ce_{0.5}Gd_{0.5}O_{1.75-\delta}$	0.499

Appendix C.

Thermogravimetric method

Figure C.1 shows a typical TG experiment. From the average of the indicated weight changes, both the OSC and the OEC are directly calculated.

$$OSC \; or \; OEC = \frac{\Delta m}{m(begin)} \frac{1}{M_{O_2}}$$ (C.1)

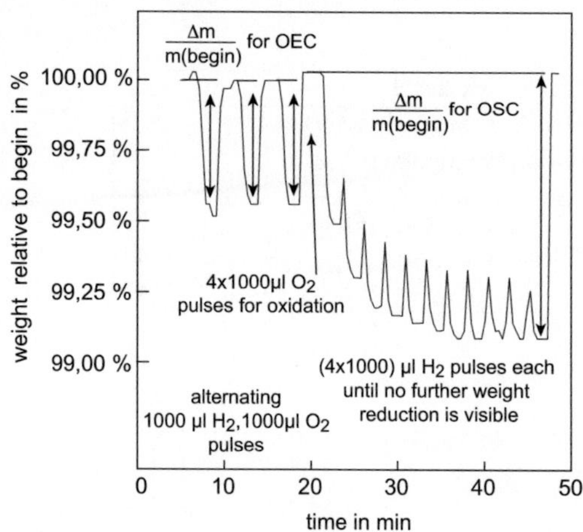

Figure C.1: Thermogravimetric method to determine OBC and OEC. CGO x_{Gd}=0.2 sample, T=800 °C

Hydrogen conversion during the experiment is low - for the CGO x_{Gd} sample, approximately 7x4000µl H_2 pulses are needed to reach the OSC. This corresponds to 1.07 μmol H_2 and a hydrogen conversion of ca. 0.5%.

134

Appendix D.

Reaction conditions

In choosing the reaction conditions, care has to be taken to exclude any possible influence of transport phenomena such as mass transport or heat transport limitations. Various methods and criteria have been developed to check for these influences. Based on several sources [94, 95, 96], an excel sheet was developed, see the following two pages and the summary below.

- the assumption of plug flow (i.e. neglecting axial dispersion) is justified (Bodenstein number Bo = 0.3)

- the bed dilution can be neglected, although the used dilution is close to the allowable maximum

- the external mass transfer can be neglected (Carberry number Ca = 4.3e-10 which is < 0.05)

- the internal mass transfer can be neglected (calculated assuming a pore diameter of 10 nm, for the agglomeated particle), Weisz modulus Ξ = 8.3e-9 which is < 0.08)

- there is no heat transfer limitation over the film surrounding the catalyst particles

- there is no temperature gradient in the catalyst particle

- there is no radial heat transfer limitation within the bed

From the above analysis can be concluded that the two parameters dilution and radial heat transfer are critical for these particular experiments - sufficient dilution is required to spread out the catalyst particles enough to avoid temperature gradients. Mass transfer limitations on the other hand are not critical in this specific case, mainly because of the lack of internal surface area and because of the small catalyst particle diameter.

input
deriv ed

conditions		
T	1073,15	K
p	1,5	bar
$^N\Phi_t$	0,398	mol/h
$^V\Phi$ (NTP)	8,92	l/h
$^V\Phi$	23,67	l/h
ρ_m	0,29	g/l = kg/m3

reaction mix		g/mol
x_{CH4}	25%	16
x_{N2}	0%	27
x_{H2O}	75%	18
M_m		17,5
cCH4	4,21	mol/m3

reactor		
m_{cat}	0,3172	g
m_{quarz}	2,1	g
ω_{Rh}	1%	
ε	0,4	
d_p	180	nm
	0,00000018	m
$d_{p,quarz}$	60	µm
	0,00006	m
d_R	7,00E-03	m
ρ_{cat}	12750	kg/m3
ρ_{quarz}	1583	kg/m3
l_R	0,0585	m
S_{BET}	4	m2/g
λ_{cat}	11,3	W/m K
λ_{quarz}	1,46	W/m K
$\lambda_{average}$	1,484	W/m K

A_R	3,85E-05	m2
V_R	2,25E-06	m3
τ	0,342	s
τ_{mod}	0,048	(g s)/ml
ρ_{bed}	140,894	kg/m3 = g/l
d_{pore}	1,57E-08	m

checking the assumption "plug flow"

Axial dispersion (calculated with quarz particle)

Bo (=P e_p) =   0,10   Valid for R e =   0,21 < 20
Bo (=P e_p) =   0,32   Perrys: Langer , 1978 & Wick e, 1973
$l_R/d_{p,A}$ (minimum required) =   269   86
$l_R/d_{p,A}$ (experimental)   =   975

radial dispersion (calculated with quarz particle)

$d_R/d_{p,quarz}$ 117   should be > 8

maximum allowable catalyst dilution

b   0,982   volume of inert material as fr   action of total
b_{max} 0,9999996

checking mass transfer limitations

External mass transfer limitation (calculated for catalyst particle)

	for cat particle	for quarz particle	
Sh = 2 + 1.1*R $e^{0.6}$ $Sc^{0.33}$ =	2,010	2,338	
(validated for 0.1 < Re =	*0,001*	*0,210*	*< 100)*
Sh = (4+1.21(R e Sc)$^{2/3}$)$^{1/2}$ =	2,001	2,001	
(validated for Re =	*0,001*	*0,210*	*< 0.1* from P erry's
and ReSc =	*0,0003*	*0,1005*	*< 10000)* for creeping flow
Mass tr ansfer coefficient (χ) =	2834,447	9,937	m^3/m^2.s
(6/d_p) =	33333333	100000	m^2/m^3-kat
$c_{CH4,bulk}$ =	4,21	4,21	mol/m^3
ΔC =	4,391E-10	4,175E-05	mol/m^3

Carberry's number (Ca) = (r $_{CH4}$)/(χ (6/d $_p$ $c_{CH4,bulk}$)=
[C(bulk)-C(surface)] / C(bulk) =   1,044E-10   9,925E-06
this must be   <   0,05

Internal mass transfer limitation (calculated for catalyst particle)

Effectiv e diffusivit y   (using a tortuosit y =   5   )
D_{eff} =   9,505E-07   m2/s

Calculation of the W eisz-Pr ater criterion
No pore-diffusion limitation if the W eisz modulus
is smaller than   0,08
Weisz modulus (Ξ) =   9,34E-09

reaction		
k'_{CH4}	70	ml/(g s)
k_{CH4}	9,863	1/s
r_{CH4}	41,49	mol/m3cat s
E_A	90	kJ/mol
ΔH_R	-206	kJ/mol
XCH4	97%	
α	1	

μ_m	0,0000359	kg/(m s)
ν_m	0,0001219	Pa s
$c_{p,m}$	0,826	J/mol K
λ_m	0,310	J/m s K
$D_{ii'}$	0,000255	m²/s

u_z	4,27E-01	m/s
u_0	1,71E-01	m/s
Re_R	24,52	
Re_p	0,0006305	
$Re_{p,R}$	0,2101664	
Sc	0,478	
D_{ax}	0,000255	$= D_{ii'} + \bar{u}^2 d_r^2/(192 D_{ii'})$
$Bo = Pe_p$	0,100481	
$Ca\ (=Da_{II})$	1,044E-10	
$D_{Knudsen,CH4}$	1,246E-05	m2/s
Pr	0,005466	

checking for heat transfer limitations

External heat transport limitation (calculated for catalyst particle)

(Heat tr ansfer limitation o ver the film surrounding the pellets)

$Nu = 2 + 1.1 R e^{0.6} Pr^{0.33}$ *[Wakao et al. (1979)]*

(analogous to 'External diffusion limitation')

Nu number = 2,00

Heat tr ansfer coefficient (a) = 3448525 W/m².K

Criterion : | $\Delta T(film)$ | = 7,435E-08 K < 5,319 K   for cat particle

Heat tr ansfer coefficient (a) = 10346 [W/m².K]

Criterion : | $\Delta T(film)$ | = 0,018 K < 5,319 K   for quarz particle

Temperature gradient within the pellet

The effect of the internal temper ature gr adient on the nett production r ate is smaller than 5% if :

1,78E-10 K < 5,319 K

Effective thermal conduct ivit y λ_{er}:

Appro ximation using the equation from Y ani & K unii (1957) and the correlation concerning the 'film thickness' from K ulkarni & Dor aisw am y (1980) yields:

(valid if 0.1 < λ_g/λ_s = 0,209 < 1)

λ_{er}^0 = 0,945 W/m K

Contribution to λ_{er}^0 via conduction through the gas phase only = 0,124 W/m K

Contribution to λ_{er}^0 via conduction through pellets and stagnant gas film = 0,821 W/m K

(The effect of heat r adiation has been neglected; this is allowed if T <1100-1400 K)

λ_{er}(dynamic) = 4,366E-05 W/m K   (De W asch &

(validated for 30 < Re = 0,210 *< 1000)*   Froment, 1972)

λ_{er} = 0,945 W/m K

Heat transfer coefficient at reactor wall (a_w):

a_w^0 = 7207,82 W/m² K

a_w = 7208,01 W/m² K   (De W asch & Froment, 1972)

(validated for 30 < Re = 0,210 *< 1000)*

Radial heat transfer limitation :

The effect of the r adial temper ature gr adient on the nett production r ate is smaller than 5% if :

| $\Delta T(r ad)$ | = 0,153 K < 5,3 K

Appendix E.

Pulse Analysis with empty Tube

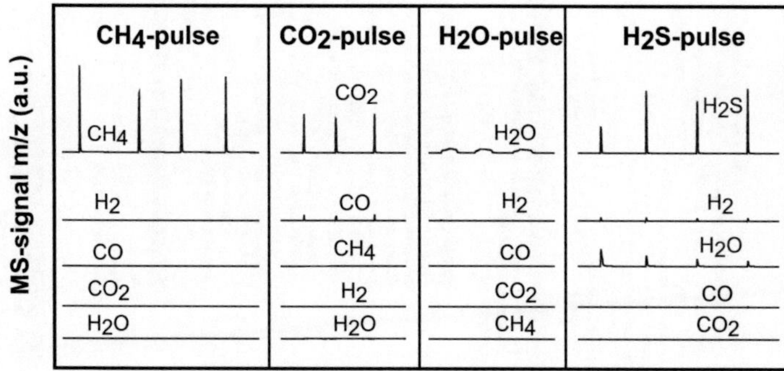

Figure E.1: Mass-spectrometer traces at the outlet of the empty reactor after injection of the indicated substance.

Appendix F.

Thermodynamic equilibrium considerations

Nitrogen conversion   If the nitrogen present in natural gas reacts, mainly ammonia, but possibly also some HCN, HNCO and other nitrogen containing compounds could be formed. Calculations with the software package ASPEN Engineering Suite revealed that NH_3 would be the main reaction product, with HCN and HNCO yields being about 200 times smaller. As expected, the amount of ammonia primarily depends on the fraction of nitrogen present in the natural gas, see Figure F.1. Due to the stoichiometry of the reaction, the ammonia content is proportional to the square root of the nitrogen content of the natural gas.

Reports on ammonia formation during hydrogen production are scarce. In the case of secondary reformers, used in the industrial production of ammonia, it was noted that only about 25% of the equilibrium value of NH_3 is obtained [97]. A likely explanation is that the catalysts typically used for steam reforming are about 4–5 orders of magnitude less active for ammonia synthesis and thus equilibration is very slow [98, 99].

In conclusion, nitrogen conversion is very small during the steam reforming of nitrogen containing natural gas. This means that the use of nitrogen as internal standard is adequate. Whether the formed levels of ammonia cause problems in downstream process equipment however remains to be investigated.

Carbon formation in equilibrium   Figure F.2 shows that the yield of elemental carbon in thermodynamic equilibrium does not approach zero until a steam to carbon ratio of $\zeta \geqslant 1.5$ is reached. For lower steam to carbon ratios, some carbon is always observed in thermodynamic equilibrium - and at $\zeta=0.5$ the expected result of 50% elemental carbon yield is obtained.

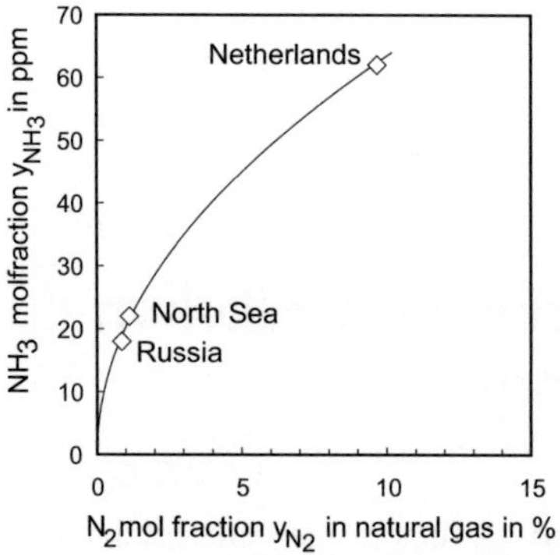

Figure F.1: Hypothetical ammonia content in the dry product gas in thermodynamic equilibrium (if it was possible to activate the nitrogen contained in natural gas by the catalysts present). T = 800 °C, p = 3 bar, ζ = 3. Line represents a square root fit, see text. Refer to Table 1.1 for the natural gas compositions.

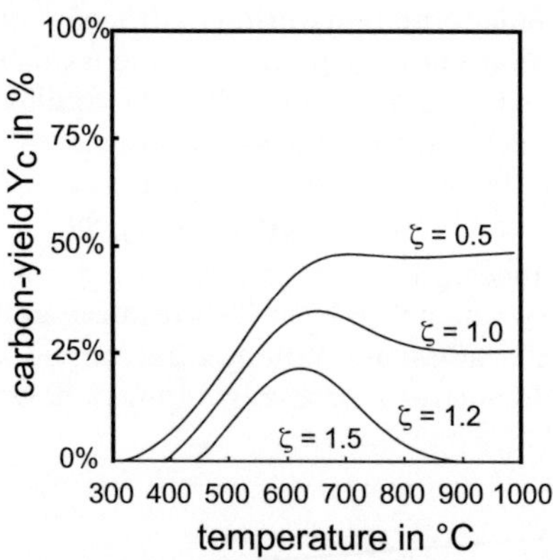

Figure F.2: Carbon yield in thermodynamic equilibrium with pure methane as feed for several steam to carbon ratios ζ. T = 800 °C and p = 3 bar.

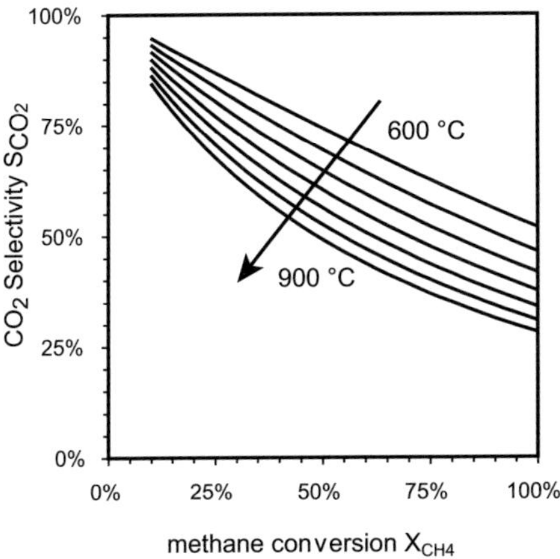

Figure F.3: Dependence of the CO_2 selectivity in equilibrium on the methane conversion at a fixed temperature. The lines shown are for a temperature range between 600 °C (highest selectivity) and 900 °C (lowest selectivity), in 50 °C steps. $\zeta=3$ and p = 1.5 bar.

Influence of methane conversion on CO_2 selectivity in equilibrium
Experimentally, the methane reforming is relatively far from equilibrium at low temperatures, see Figure 3.17. In those cases, the gas composition is quite different in the experiment compared to a calculation in which equilibrium for both the methane reforming and the shift reaction is assumed. As the methane reforming yields 3 H_2 molecules, a higher methane conversion pushes the water-gas shift reaction to the left side in Equation 3.4, i.e. to lower CO_2 selectivities. Reversely, a lower conversion as in equilibrium, as observed experimentally, would allow for higher CO_2 selectivities than expected by an equilibrium calculation with both reactions in equilibrium. This is indeed seen experimentally - see Figure 3.17.

Figure F.3 shows the dependance between the methane conversion, temperature and the CO_2 selectivity in equilibrium.

Appendix G.

Error analysis for the TOF calculation

In principle, the uncertainty of a parameter which is calculated from several measurements is composed of the individual measurements times the "impact" that measurement has on the parameter. Mathematically this is expressed by Gauss' law of propagation of uncertainty:

$$\Delta TOF = \sum_{i=1}^{n} \frac{dTOF}{dx} \Delta x \qquad (G.1)$$

With x being the individual measurement.

Rather than deriving the individual $\frac{dTOF}{dx}$ factors from Equation 2.22, a numerical approach was taken. Table G.1 summarizes the numerical values

Table G.1: Elemental analysis of the samples

parameter	$\frac{dTOF}{dx}$	including all errors		excluding systematic errors	
		Δx	$\frac{dTOF}{dx}\Delta x$	Δx	$\frac{dTOF}{dx}\Delta x$
y_{CH_4}	207	0.01	2	0	0
m_{cat}	173	0.1 mg	0.017	0.1 mg	0.017
Y_{Rh}	5232	$1\,10^{-6}$	0.005	$1\,10^{-6}$	0.005
$V\Phi_{total}$	5,8	$0.5\,\frac{l}{h}$	2.9	0	0
X_{CH_4}	577	2.0%	11.5	1.0%	5.8
D_{Rh}	487	0.5%	2.4	0.5%	2.4
Δ TOF			18.9		8.2

Appendix H.

Rawmaterial cost for the catalyst

Table H.1: Rhodium and Nickel Catalyst aas well as traditional ZnO adsorbent raw materials prices (2006 prices, [100, 101, 102, 103, 104, 105])

Material	Price in \$/t
Nickel	24
Aluminum	2 557
Magnesia (MgO)	540
Zinc	3 274
Rhodium	146 000 000
Ceria (CeO$_2$)	40 000
Gadolinia (Gd$_2$O$_3$)	140 000

Lebenslauf

Persönliche Daten

Name	Ulrich Markus Hinrich Hennings
Geburtstag	04. April 1975
Geburtsort	Rottweil am Neckar
Familienstand	verheiratet

Schulausbildung

1981–1985	Eichendorff-Grundschule in Rottweil
1985–1994	Leibniz-Gymnasium in Rottweil
17. Juni 1994	Allgemeine Hochschulreife

Wehrdienst

1994–1995	1./Feldjägerbatallion 750, Stetten a.k.M.

Studium

1995–1998	Studium des Chemieingenieurwesens an der Universität Karlsruhe(TH)
1998–2000	Studium im Fach „chemical engineering" an der University of Wisconsin, Madison, USA
20. Mai 2001	Abschluß mit dem Titel „Master of Science-Chemical Engineering"

beruflicher Werdegang

04.2001–12.2001	wissenschaftlicher Angestellter an der DVGW Forschungstelle am Engler-Bunte-Institut der Universität Karlsruhe (TH)
01.2002–06.2006	wissenschaftlicher Angestellter am Engler-Bunte-Institut der Universität Karlsruhe (TH), Lehrstuhl für Chemie und Technik von Erdöl, Erdgas und Kohle
seit 11.2006	„Process Technologist" bei Shell Global Solutions, Amsterdam

Ende des Tunnels erkennen zu lassen. Ich war schon drauf und dran, mich darin zu verlaufen.

Der großartigen Almut für stete Unterstützung und lange Gespräche, die schließlich doch zum Ziel führten.

Vielen Kolleginnen und Kollegen für Infos, Infos, Infos … und eine schöne gemeinsame Zeit.

Margit und Rainer für schicksalhafte Momente und die herzliche Mithilfe.

Dr. Rainer Schöttle für konzeptionelle Beratung und Bearbeitung der allerersten Güteklasse.

Herrn Landesgeschäftsführer Leonhard Stärk von der Landesgeschäftsstelle des Bayerischen Roten Kreuzes in München für den unablässigen Einsatz für seine Sache.

Danksagung

Folgenden Personen, die wesentlichen Anteil daran hatten, dass dieses Buch überhaupt entstand, möchte ich meinen Dank aussprechen:

Dem lieben Annerl für ihre stete Unterstützung und selbstlose Mithilfe in dieser schweren, aber auch kreativen Zeit.

Meinem Sohn Niki für die Idee und den sprühenden Esprit seines Entwurfs für den Einband.

Holger Bogen (Clever Cartoon) in Berlin für die professionelle Ausarbeitung des Einbands und seine kompetente Beratung. Ganz große Klasse!

Patti in Ljubljana für die gekonnten Illustrationen und viel zu vielen gemeinsam vertilgten Cevapcici.

Der Brauerei Unertl in Haag (Obb.) für ihr eingebungsvolles dunkles Weißbier, den Erinnerungstrunk.

Martin Wehrle, dem Verfasser der »Ich arbeite in einem Irrenhaus«-Bücher, für daraus gewonnene Erleuchtung, seinen überaus freundlichen Zuspruch und das unerwartet vermittelte Quantum Elan, um mich das Licht am